Adam Storey Farrar

Science in theology

Sermons preached in St. Mary's, Oxford, before the university

Adam Storey Farrar

Science in theology
Sermons preached in St. Mary's, Oxford, before the university

ISBN/EAN: 9783337113414

Printed in Europe, USA, Canada, Australia, Japan

Cover: Foto ©berggeist007 / pixelio.de

More available books at **www.hansebooks.com**

SERMONS,

PREACHED BEFORE THE UNIVERSITY OF OXFORD.

SCIENCE IN THEOLOGY.

SERMONS

PREACHED IN ST. MARY'S, OXFORD,

BEFORE THE UNIVERSITY.

BY

ADAM S. FARRAR, M.A., F.G.S., F.R.A.S.,

MICHEL FELLOW OF QUEEN'S COLLEGE, OXFORD; LATE ONE OF THE SELECT
PREACHERS TO THE UNIVERSITY; AND PREACHER AT THE
CHAPEL ROYAL, WHITEHALL.

PHILADELPHIA:
SMITH, ENGLISH & CO.,
No. 23 NORTH SIXTH STREET.
NEW YORK: SHELDON & CO.
BOSTON: GOULD & LINCOLN.
1860.

PREFACE.

The title, "Science in Theology," is intended to express the writer's purpose in the composition of the following Sermons, viz., to bring some of the discoveries and methods of the Physical and Moral Sciences to bear upon theoretic questions of Theology.

The history of the growth of systematic theology on the one hand, and of religious scepticism on the other, exhibits marked traces of the constant presence of an element which may be called Science in Theology. From the time that Theology first arose out of Religion, the speculative theory out of the practical art, it has never failed to receive a tinge from the condition of general knowledge existing, and the methods for the investigation of truth prevalent, in each particular age. Itself a kind of science,—so far as systematic arrangement of principles can constitute science,—it has shared the fate of the other sciences; it has been compelled to take its place among them, and has met with opposition, or has received illustration from them. Its history is marked by epochs of criticism or of scepticism, in which it has had to submit to the investigations of co-ordinate bodies of Physical or Mental Philosophy, sometimes refuting them, sometimes borrowing from them, at other times surrendering to them. In each of these epochs the difficulties presented have been grounded in some form of Science or Philosophy which has been brought to bear upon Christian Theology; in each of them the restoration or the perpetuation of Christian belief has depended upon the readjustment of the new form of thought with the claims of pre-existent religious dogmas. The battle has been metaphysical or scientific, not strictly theological. It has been fought in reference to the premises from which the sceptics or critics have started, not to the conclusions at which they have arrived.

In the early centuries, for example, Theology received a tinge from contact with the allegorizing philosophy of Alexandria, which expressed itself in the writings of Origen. That learned man could not lay aside his favorite habits of thought, but strove to adjust Christian speculations to them. During the two centuries which followed his time, Theology came into conflict with the Neo-Platonic philosophy,* and in the conflict came forth victorious from the first great historical epoch of scepticism. In the middle ages it encountered a new danger, a second crisis, from the criticism of Nominalists like Abelard, in the University of Paris;† and it received a new adjustment with the existing state of thought through means of the logical arrangements of the Schoolmen, such as Anselm and Aquinas.‡ At the era of the Renaissance it encountered new difficulties in being brought into contact with the wider knowledge which Providence, from time to time, disclosed to mankind—difficulties which have not, like the more ancient ones previously noticed, quickly expired, but have left their effects to the present day. The sacred books were then, for the first time, exposed to the criticism of great scholars and editors, and alarm was excited by the discovery of variety of readings in the text. Received dog-

---

* See Sermon V (pp. 142-146) of this volume.

† See Sir J. Stephen's "Lectures on the History of France" (vol. ii, first Lecture on the Power of the Pen).

‡ See Sermon VI, pp. 171, 172.

mas also were submitted to the acute controversies generated by the Reformation, and while undergoing revision from that movement the terminology in which they were expressed became stereotyped in the mode that might be expected from an era of religious struggle.* In the early part of the seventeenth century the discoveries in Physical Science also began to unveil new truths to the minds of the orthodox. Metaphysical Science likewise,

---

* It was an age when theologians had not time for careful thought, even if the state of scholarship had been sufficiently advanced to supply them with the materials for it. This, however, was not the case. In the world of scholars there were indeed giants in those days—Erasmus, Budæus, the Stephenses, the Scaligers, &c.,—but their attention was devoted mainly to *words*. The scholars of the seventeenth century applied themselves rather to *things*. On the foundation thus prepared, the German scholars of the last and present century have been able to found accurate sciences of language and of criticism. While, therefore, we justly render all honor to the noble efforts of the theologians of the era of the Reformation, we ought not to suppose that they, with their imperfect attainments, were infallible interpreters of the sacred Scriptures, nor to allow their views to be an impediment to the theological progress which Providence is forcing on the world by the advance in knowledge, to a portion of which allusion has been made. The stand-point of the sixteenth century had its value; it was a noble protest against mediæval Christianity, an epoch in spiritual emancipation; but it was not the same as the stand-point of the first century, and for this reason, as well as for the others just named, it is not sufficiently broad and simple and learned to be the stand-point of the nineteenth.

in opening up investigations into the origin of knowledge, led through the spread of Sensationalism to a new, the third, epoch of religious scepticism, which is generally identified with the name of Voltaire and the contemporary philosophers of France. It is unnecessary to point out the intellectual and moral means by which Theological Science outlived this new crisis. The battle was again fought on Scientific ground, not on Theological. Lastly, the convergence of different lines of thought in the present day; of the Intellectualism of Germany* and the Positivism of France; of religious dogmatism and scientific scepticism; and the existence of apparent discrepancies between Theology and the Sciences, are producing a fresh era of criticism, a fresh crisis of doubt. Theology must again listen to secular discoveries, must refute them or readjust its doctrines and its methods to them; and the humblest attempts, made without sophistry, in an honest and loving temper, to aid in such a desirable result, must surely be useful.

The history thus given of Science in Theology, *i.e.* of the relation which Science, Physical and

---

* This term is intended to include the Spiritualist tendencies of the followers of Schelling, as well as the Rationalist followers of Kant and Hegel.

Mental, has held to Theology, will explain clearly the meaning of the writer of these pages. The following sermons can only be regarded as detached contributions, aiming to show the mode in which the Theology of the present day may incorporate the irrefragable discoveries of modern science into its own system. They will have performed their office if (with the assistance of the hints offered in the foot-notes) they are the means, by God's blessing, of suggesting materials for reflection to thoughtful and religious minds. They are published in deference to the wishes of very many persons, who made the request at the times when they were respectively preached.*

The topic thus indicated offers abundant field for investigation to those theological students who desire to aid in removing the difficulties which many educated men in the present day feel in reference to our holy religion. It may possibly assist their studies to specify the six following branches of in-

---

* They are arranged according to the internal connection of subject rather than the chronological order of delivery. In order to complete the series, one sermon has been added which was preached at Whitehall, but some parts of it were contained in a Sermon (not published) preached before the University.

quiry which still, perhaps, require farther treatment:*—

First: The relations of metaphysical science to religion demand an investigation of the limits which the structure of the human mind imposes in reference to theological speculation.

Secondly: A comparison is needed between the statements of the sacred inspired books of the Jews and the modern discoveries of comparative Philology and Mythology, and of historic data derivable from recently excavated inscriptions.

Thirdly: A discussion of the relations of the Physical Sciences to the evidences of Natural Religion and to the doctrines and inspiration of Revealed.

Fourthly: An examination of the various books of the Holy Scriptures as works of literature of the ages in which the authors of them lived.†

---

* The above remarks, with some of those which immediately follow, express the substance of an unpublished Sermon, on "The Epochs of Religious Scepticism," preached by the Author before the University, on St. Thomas's Day, 1855. It is for the use of those who requested the publication of it that they have been introduced in this preface.

† In a manner, for example, analogous to C. O. Müller's "History of the early Literature of Greece." The literature of the Jews, their early poetry, and eloquence, and chronicles, and proverbs, possesses, like that of Greece, the value of being an original and (so to say) in-

Fifthly: A history of the Science of Theology and of the growth of Theological opinions.

Sixthly: A candid, reverent, and loving examination of the light which these various branches of inquiry throw on the proof, nature, and limits of Inspiration.

It is gratifying to think that many works of great importance have recently appeared in reference to most of these lines of inquiry.* The writer may

---

digenous literature. It has, therefore, the special claims for psychological study which only belong to those literatures which we have reason to regard as original utterances of the human intelligence, and expressions of national life and modes of thought. Such an investigation is a necessary preliminary to a right comprehension of the distinctive, inspired, and supernatural element which enters into the Jewish literature.

* The first of these subjects, for example, has been recently treated by Mr. Mansel in his Bampton Lectures, and by Mr. Maurice in his reply, "What is Revelation?" the second by Mr. Rawlinson, in the Bampton Lectures for 1859; the third by Prof. Baden Powell; the fourth by Ewald; the fifth partially by Baur and Bishop Hampden; (also the instructive writings of Professors Stanley and Jowett, and the matchless sermons of the lamented F. W. Robertson supply many very valuable hints in reference both to this topic and to the one last mentioned); the sixth by Dr. Lee, by Mr. Morell in his "Philosophy of Religion," and by Dean Harvey Goodwin in two of his Hulsean Lectures. The above works are only a few of those which have appeared. A rich mine, almost unworked, of novel and most suggestive thought exists in the modern theological literature of Germany for any one who can separate the pure metal from the dross. In naming these works

perhaps be permitted to state that he hopes hereafter to discuss the third and sixth of them, towards which his studies, conducted in the retirement of an Oxford cloister, have through several years been made to converge.

the writer does not by any means intend indiscriminate praise or acquiescence. He bestows his praises as a student, not as a theologian; the difference being that a student usually praises a work if he finds it suggestive of thought and instruction; while a theologian too often withholds his commendations if he finds the sentiments disagree with his preconceived opinions, *i. e.* (to use Lord Bacon's caustic expression) with the idols of his own den.

TABLE OF CONTENTS.

SERMON I.

THE GRADUAL DISCOVERY OF THE DIVINE ATTRIBUTES THROUGH SCRIPTURE AND SCIENCE.

ISAIAH 57 : 15.

PAGE

The historic method of studying the development of theological opinions applied to trace the knowledge of the Divine attributes derivable through—I. *Reason*, II. *Revelation*, III. *Science.*—I. Distinction between the origin of the idea of God and the origin of the idea of His attributes.—The three theories on the former; the successive efforts of the Greek philosophers (Thales, Anaxagoras, Plato, Aristotle) in reference to the latter.—II. Views of God's attributes gradually unfolded in the successive stages of Revelation, as shown in the thoughts of Moses, David, Isaiah, and in Christianity.—III. Science regarded as a new and additional revelation, given without miracle, through uninspired genius.—Knowledge of the Divine attributes unfolded by means of it, *e. g.* through Astronomy, through the telescope and microscope, through Geology, Mathematics, Moral Science.—Inferences, ...   25

NOTE, On the Theological Ideas of some early Greek Philosophers,   51

SERMON II.

DIVINE PROVIDENCE IN GENERAL LAWS.

Acts 17 : 28.

PAGE

Description of Athens in the time of St Paul, and of the Epicurean and Stoic theories of Providence.—Statement of causes which have produced sceptical theories on Providence in succeeding ages, viz.: the speculations of Metaphysicians and the real discoveries of general laws through Natural Science and through Statistics.—The object of the Sermon stated to be (I) the discussion of the scientific discoveries in reference to the invariability of Nature's laws, and (II) the reconciliation of such a view of Providence with the idea of benevolence in God. —I. Illustrations of the fixity of Nature's laws in the example of, (α) earthquakes (e. g. that of Lisbon) and, (β) in colliery explosions.—Proof that such catastrophes are not explicable as judgments for sin, both from our Lord's teaching and from the Geological discovery of the existence of pain and death antecedent to human history.—Further exemplification of the invariability of Nature's laws in (γ) the permission of historical and political catastrophes like the Indian rebellion of 1857.—Theory of such events.—II. Reconciliation of the Divine government by general laws, which has thus been proved, with the idea of benevolence in God's character.—The "greatest happiness" principle applied to this subject.—Illustration from Lagrange's discoveries in Mathematical Astronomy.—Remarks on the apparently contradictory doctrine of Special Providence taught in Scripture as equally real with the Providence taught in science, 54

Note, On Special Providence, ... 80

SERMON III.

DIVINE BENEVOLENCE IN THE ECONOMY OF PAIN.

GENESIS 47 : 8, 9.

Description of the Egypt of the age of Jacob, and of Jacob's melancholy view of human life.—Subject of the Sermon stated to be the Theory of Pain, *i. e.* the reconciliation of the existence of pain with the attribute of benevolence in the Creator (I) in the purpose of its administration, and (II) in the remedies provided for its diminution.—I. Discussion of, (α) the pain which can be clearly shown to be corrective, not retributive; with illustrations on this subject from the Geological discovery that pain existed antecedently to human history, and therefore antecedently to human sin. (Examination, in a foot-note, of the alleged discrepancy between this discovery and the teaching of Holy Scripture.) (β) The pain which arises from accidents and from the operation of general laws, which is equally proved in the large scale to be compatible with the Divine benevolence.— II. Further proofs of God's kindness shown in the remedies provided by Providence, (1) in the growth of medical science, (2) in the growth of public opinion and sympathy, (3) in the philanthropy awakened by civilization and Christianity.—Lessons recently taught on the importance of attention to the secular aspect of religion, ... 82

NOTE, On the Evidence of Geology, designed to refute some objections taken against the science, ... 103

SERMON IV.

JEWISH INTERPRETATION OF PROPHECY.

Isaiah 6 : 9.

PAGE

The pertinacity of the Jewish national character shown to be a cause which has impeded attempts made for their civilization or conversion.—Object of the Sermon explained to be, (I) a sketch of the history of Jewish *uninspired* theological literature, and (II) the statement of the principal lines of argument used in the discussion between Jews and Christians.—I. Their theological literature studied in three epochs:—1. From their captivity to the Christian era.—Effects of the captivity on the Jewish national character and position, with the causes which created that branch of literature called the Targums; 2. From the third to the eighth century, A. D.—Sketch of the two centres of Jewish life in Galilee and Mesopotamia.—The school of Biblical criticism creates the Masora, and that of Biblical interpretation produces the two parts of the Talmud, the Mishna and the Gemara; 3. From the tenth to the fifteenth century in Spain.—The literary condition of the Jews in Spain. Schools of theological interpretation, *e. g.* Jarchi, Aben Ezra, Kimchi, and of philosophical theology, *e. g.* Maimonides.—Subsequent writers, Abarbanel and Rabbin Isaac.—II. The controversy between Jew and Christian shown to depend upon interpretation of prophecies.—The tests of true interpretation resolve themselves into a question of circumstantial evidence.—Three lines of objection urged by the Jews against Christianity considered and refuted:—1. The historical and popular one, that Christianity was rejected by contemporary investigators; 2. The philosophical one of Maimonides, that an incarnation of God is impossible, and contrary to the analogy of the old Hebrew religion; 3. The critical one, of the Spanish school of Jewish interpreters, that the Messianic prophecies are misinterpreted by Christians.—Remarks on the need of a more correct system in prophetic interpretation.—Concluding inferences, ... 106

SERMON V.

THE DOCTRINE OF THE HOLY TRINITY.

EPHESIANS 2 : 18.

Peculiarity of the doctrine, among all others in Christianity, in depending for its proof entirely on revelation.—The practical statements of Scripture in relation to it contrasted with the theoretic terms of Creeds.—I. History of controversy in reference to the doctrine :—1. In the Greek university of Alexandria.—Sketch of the Neo-Platonic philosophy, first as a metaphysical, secondly a political, thirdly a logical movement; and its relation to the doctrine of the Trinity; 2. In the rise of Socinianism in the sixteenth century; 3. In the philosophical systems of modern Germany and of Coleridge.—II. The doctrine which is really to be believed as the residuum after the theories which have incrusted the inspired teaching are removed.—The idea of *personality* shown to be derived from human analogy, but to convey a true fact.—The influence of science shown in creating a moral disposition to believe mystery if it rest on evidence.—Practical inferences, ... 138

SERMON VI.

THE ATONEMENT.

Mark 9 : 2.

PAGE

The Transfiguration studied:—1. In its geographical scene; 2. In its moral meaning.—The fact that our Saviour's teaching entirely changed from this event (for previously to it he had not proclaimed the fact that he was to *suffer*) used as a proof that the Transfiguration had a real meaning in reference to our Lord, and was not merely a parable acted to instruct the disciples.—(Illustrations of similar mystery beyond their moral meaning to us in our Lord's Temptation and Agony.)—Statement of the modern theories on the purpose of our Lord's suffering, and also of the Apostolic teaching in reference to it.—The subject of the Atonement the chief purpose of this Sermon.—I. An historic sketch of the successive theories on the Atonement from the Apostolic times:—(α) the Patristic to A.D. 1000, which taught that it was to ransom man from the Devil (the view, *e.g.* of Gregory the Great); (β) the Scholastic (A.D. 1000—1500), that it was a satisfaction rendered to a broken law, either by the life of Christ (Anselm's view) or the death of Christ (Aquinas's); (γ) the Protestant view (of the Reformers and Grotius, A.D. 1500 —1700), that it was a satisfaction for sin viewed under the *corrective*, as distinct from the *retributive*, theory of punishment; (δ) the view of modern German philosophers, that it was only to reconcile man to God, not God to man.—II. Refutation of the first three views, as going beyond inspired teaching, and constructed in forgetfulness of the proofs which Physical Science gives, of the poverty of human intelligence, and Mental science of the extent to which analogy is the medium of revealed truth.—III. Hints for the refutation of the fourth view, as falling short of the truth.—Criticism on other modes adopted for refuting this fourth view, such as the ordinary one and Mr. Mansel's.—Another suggested, resting on the idea of *guilt* and the fact of *sacrifice*.—Conclusion as to the reality of Christ's death being a true but undiscoverable means of reconciling (so to say) God to man, as well as man to God, ... 157

SERMON VII.

LAWS IN THE LIFE SPIRITUAL.

2 Timothy 4 : 7.

Statement of the three forms of human life, the practical, the intellectual, and the spiritual.—St. Paul's character studied as an harmonious embodiment of them all; and his influence estimated at different periods of church history in awakening theological speculation and stimulating religious effort.—The religious life studied as follows:—I. Is it subject to fixed laws? Three opinions stated; (α) that which resolves it into ordinary processes of moral psychology (Butler's view); (β) that which makes it dependent on Divine election (Calvin's); (γ) that which makes it exist in a faculty transcending consciousness (Schleiermacher's).—II. Are its laws discoverable?—III. By what means? By induction from inspired and uninspired religious biography.—Enumeration of tests applicable to insure correctness in such discovery.—IV. Laws of religious life derivable from St. Paul's life, in reference to conversion, assurance, salvation in death, ... 184

SERMON VIII.

THE GIFTS OF THE HOLY GHOST.

John 14 : 16.

PAGE

The miracle of the descent of the Spirit compared with the miracles of Creation and of Redemption.—Four great gifts of the Spirit to be studied:—I. Miracle.—The comparison of Christianity with Boodhism and Mahometanism an evidence of the reality of ancient Christian miracles.—Reason of their disappearance.—II. Inspiration.—How far it was, like Miracle, a temporary gift.—III. Holiness—studied in apostolic Christians and in the history of the religious consciousness—shown to be a gift for all time.—IV. Supernatural religious usefulness.—In what sense perpetual.—Illustrations from history of religious reawakenings in later times, *e. g.* the ministry of Francis of Assisi, Methodism, the Rev. C. Simeon, &c., . . . 208

SERMON IX.

PROVIDENCE IN POLITICAL REVOLUTIONS.

Proverbs 16 : 4.

Discussion of the Oriental (apparently fatalistic) modes of expression in Scripture.—Instances of good brought by Providence out of evil, (α) in Jewish, (β) in general, history.—Application of this principle to the English Revolution in the reign of Charles I, in the form of a proof that Providence overrules national convulsions to ultimate good by a law which is impressed on society.

PAGE

—The relation of this argument shown to the evidence of beneficence in the Divine character.—1. Political convulsions overruled for good in producing the material welfare of man, the advancement of liberty.—The causes of revolution explained, and their application to the example of the English Revolution by detailed reference to the history.—2. Their influence on the moral discipline and instruction of nations.—Concluding inferences, ... 228

NOTE, On the Scene of the Execution of Charles I, . . 249

SERMONS.

SERMON I.

THE GRADUAL DISCOVERY OF THE DIVINE ATTRIBUTES THROUGH SCRIPTURE AND SCIENCE.

(PREACHED BEFORE THE UNIVERSITY, MARCH 4, 1855.)

ISAIAH 57 : 15.

"Thus saith the high and lofty One that inhabiteth eternity, whose name is Holy; I dwell in the high and holy place, with him also that is of a contrite and humble spirit, to revive the spirit of the humble, and to revive the heart of the contrite ones."

IN these words the prophet combines the majesty of God with His mercy, the magnificence of His infinite power with the tenderness of His unbounded condescension. It is this combination of attributes which men are apt to regard as almost incredible,—that He who inhabiteth eternity can yet dwell with a being so inconsiderable as man. If you tell them of the greatness of God's nature, they think it impossible that He can concern Himself with reviving the spirit of the humble; or inform them that He stoops to dwell in the heart of the contrite, they can hardly imagine that He is the high and lofty One who inhabiteth eternity. If the Deity be exhibited as busied with what they deem insignificant, their inference is that He cannot be attentive to what is vast; or if He be represented as occupied with what is great, there is an immediate apprehension that the minute must escape His observation.

Nor is this disposition to separate the properties which the prophet combines more observable than the variation which it has undergone in different ages and under different circumstances. The progress, or alteration, which takes place in human opinions and belief is, though less observed, as real as that which occurs in the world of events. A rise and fall of empires is as truly going forward in the intellectual as in the historical world. Nor can there be a more instructive mode of viewing a truth than by showing the fluctuation of human thought in relation to it.* Thus, in reference to the present doctrine, there have been ages of the world when those who held fast their faith (imperfect though it was) in Providence, have failed to ascend to the idea of a Being of infinite greatness; while those, on the other hand, on whose minds speculation had forced the conviction of man's unmeasured inferiority, have doubted that an unwearied Providence could be engaged on his behalf.† And, in the present age, it is often found that those who believe in a special Providence on the authority

---

* It is probable that the sole permanent contribution to knowledge which the philosophy of Hegel will be found to have made, will be in its creation of the *historic method* of studying opinions,—a method which was in his system a necessity arising from his point of view, but which is worthy of imitation by those who differ from his motives and principles.

† These doubts marked the philosophy of the early part of the last century which followed on the great discoveries which Newton had made in the preceding age. Pope's "Essay on Man" gives expression to such doubts, borrowed probably from Bolingbroke.

of Scripture, do not understand that general Providence which is established by the evidence of science; or, that conversely insisting upon the administration of the universe by a system of general laws, they fail to reconcile it with the revealed account of God's interposition by miracles and special Providence.

It will not, therefore, I should hope, be an unprofitable employment if we trace by what means and with what degree of increasing evidence the two doctrines of the greatness of the Divine attributes and His condescending mercy have been made known to man; and, afterwards, attempt briefly to deduce from the subject lessons for our religious improvement.

There are three means by which men have been made the recipients of ideas concerning the Divine Being, and the relations which He sustains towards us, viz.: REASON, REVELATION, and SCIENCE.* Let us try to discover what assistance these respective sources have contributed towards the comprehension of the two doctrines which we are studying.

1. When we point to Reason as one source from which man has learned the greatness or the condescension of God, we do not intend to express any opinion on the question whether the first idea of a Divine Being was extracted by the mere light of natural reason, or was a direct gift of revelation. The question which concerns us is not as to the means by which men first came to learn the idea of God, but rather the process through which, when that idea was already present, they first attained to a conception of His infinity. The inquiry into the origin of the idea will

---

* By the term "Reason," is here meant metaphysical speculation; and by "Science," modern inductive discoveries.

always be a matter of uncertainty, inasmuch as it belongs to that class of questions which concern the first origin of things, such, for example, as those which treat of the origin of matter, or life, or language, or society,—questions which not only relate to facts anterior to the dawn of history, but for the explanation of which man hardly possesses the faculties.* Hence there will doubtless always be distinct opinions on the inquiry. Some,† who conceive that man is at his birth intellectually a blank, will suppose that he learns the idea gradually by experience and observation; others,‡ supposing themselves unable by such a process to explain how the world became furnished, at the very earliest and most rudimentary periods of its history, with an idea which would appear rather to be the discovery (if discovery at all) of cultivated powers, have thought that the human mind is not ushered into the world a blank, but is furnished with a small stock, as it were, of rational principles, from which germinates the variety of knowledge which forms the mental inheritance of man. Others,§ again, on the ground both of philosophy and of Scripture, decline to resign their belief that the ideas of God and of duty were originally imparted by direct revelation from heaven; conceiving that as a savage race, when sunk below a certain stage of barbarism, can never rescue itself from its degradation without being raised out of it by means of external agency,

---

* The distinction here designed is neatly expressed by Chalmers (Introduction to his "Bridgewater Treatise") as that which exists between the "collocations of matter" and the "laws of nature." Modern science is content to restrict itself to the latter of these two branches of inquiry, and to leave the former to metaphysicians.

† The sensational school of philosophers.

‡ The idealist school.

§ Locke may be perhaps taken as a type of this view ("Essay on Human Understanding," b. iv).

so the race of man could never have taken the first step in religious knowledge if Providence had not communicated to it the rudiments.

But whichever of these views you may adopt as to the origin of the idea of God, it must ever be an important inquiry to discover how much mankind can learn or have learned of God, and of the relation which He bears to us, without the aid of further revelation than the primitive one which lighteth every man that cometh into the world. The study of natural religion (as such inquiry is called) not only thus affords a valuable and independent support to the truths which revelation asserts, but also enables us to see what were the limits within which that which may be known of God had been manifested to men, which an Apostle* thought left them without excuse. Above all, the contrast of that darkness amid which they attempted to grope their way to truth, will prepare us for seeing the noonday brightness which has been thrown over these doctrines in the later ages of the world.

It fortunately happens that we need not enter into any speculations on this question, for history supplies an instance of a nation where a natural religion was actually created by the light of reason; or where, possibly, to speak more truly, the fragment of truth, which formed the last relic of an earlier faith, was matured and developed into greater purity. The philosophers of Greece worked out a natural theology.

It was hardly, indeed, to be expected that such a subject would be overlooked by that people, which stands conspicuous above all others of ancient or modern time for the natural gift of commanding intellectual faculties, and the

---

* Rom. 1 : 20.

power to appreciate the true and the beautiful; whose influence, unlike that of other nations, has been greater since its decline than when flourishing in greatness, and whose writers, as long as mankind can appreciate through the medium of an unrivalled language, the brilliancy of unrivalled thoughts, will continue to hold (as they have ever held) the same position in relation to civilization which the Jewish nation has sustained in relation to the growth of religion. Yet the very process by which the writers of that acute people discovered their natural theology, no less than the results at which they arrived by it, afford the best evidence both of the difficulty of the inquiry and the value of a Divine revelation. By successive steps, and after long intervals of time, one and another notion was added to develop into a clear conception their idea of the Divine Being. We cannot, perhaps, find a more comprehensive view of the opinions at which they had arrived in the maturity of thought, than that which is contained in a brief passage of one of their most brilliant writers, who, in the fourth century before Christ, unfolded his own view of the Divine Being, and borrowed from his predecessors those elements which they had respectively supplied. "The great first cause," Plato teaches, "is endowed with life, with intelligence, with goodness."* Of the properties which are stated in this noble conception of the Divine Being, the last alone is his own discovery. When he thus wrote two centuries had elapsed from the time when the question had been first proposed, and it will be instructive to remark the mode in which the inquiry had been from

* ἔμψυχος, ἔννους, ἀγαθός;. (Plato, Phileb. p. 30.) To this enumeration ought to be added the attributes given in Plato's "Republic" (b. ii), viz., "the cause of goodness," "unchanged" by external agencies, "unchanging" from internal.

time to time suspended, and the periods at which the elements in the answer of it were respectively furnished.*

It was towards the end of the seventh century, before the Christian era, at a time when continental Greece had not emerged from her early barbarism, that the Greek colonies which fringed the shores of Asia Minor and of Southern Italy attained that state of material civilization in which superior minds are able to devote their leisure to speculation.† It was then that the series of inquiries commenced, often daring, more often ineffectual, into the causes of the physical and moral world which has formed the noblest occupation of the mind from that time to the present. We are apt, as we look back upon that period, to undervalue the step which society then took. The discoveries, as they were called, which were then made, seem to us so rudimentary or so unreal, that we are in danger of misunderstanding the reach of thought which was necessary to attain even to them.‡ It was then that society was passing through one of those changes which mark the intellectual growth of every nation. It was awakening

---

* The writer of this Sermon wishes to state, that more careful study of Greek philosophy, during the four years since the Sermon was preached, has convinced him that he has here over-estimated the amount of theological speculation which existed among the early Greek thinkers. He accordingly now adds a note at the end of the Sermon, to correct some points stated in the text.

† For the study of this flourishing period in the Greek colonies, see Grote's "History of Greece," iii, ch. 22; and Thirlwall, ii, ch. 12.

‡ The advance of thought which Thales shows beyond his predecessors may be illustrated by the transition which Comte points out from the theological to the metaphysical, of the three stages through which he supposed that knowledge passed. Thales attained the idea of *cause* and of *unity* of cause. Measured by the fetish-like conception of power which existed antecedently, some progress may be seen even in these crude notions.

from a state of blind superstition to one of reflection. We may understand it by comparing it with the resurrection of the mind of Europe in the eleventh century of the Christian era, from the intellectual death which had passed alike over civilization and religion after the fall of the Roman Empire. The contrast is not more marked between the state of Europe in those ages, when infantile superstitions, such as the legends of the saints, were at once the object of popular faith and the medium of religious education, and that manly state of sentiment which was aroused in Europe by the growth of the scholastic philosophy,* than between the condition of Greek life which preceded and which followed the speculations, crude though they were, of the early thinkers of Ionia. For those men threw aside the polytheism of their early education, and learned to regard the cause of all things as one; and, in doing so, they took a step as it were to a conception of the greatness of the Divine Being. Yet the difficulties which beset their way will be seen from the fact that the leader of that band, Thales, in spite of possessing on the one hand the ideas of God as a first great cause, and also the idea of the unity of this cause, was unable to combine the two so as to infer the Divine Personality. Strange indeed, yet so it was, that men could understand the unity—nay, the infinity—of the first cause before they were able to discover that those attributes imply the existence of a personal mind.

* The beneficial influence here attributed to scholasticism is only in its value as a discipline. It taught little real truth; but it exercised the faculties. Its character has been well described as being "the noblest philosophy ever ruined for want of *matter*, as the cotemporary Troubadour poetry was the noblest poetry ever ruined for want of *form*." The date of the eleventh century above given is only intended to indicate the dawn of intellectual reillumination. The influence of scholasticism was in the twelfth and two following centuries.

The inquiry, as if in marvellous confirmation of the difficulty with which the reason ascends to a true idea of God, was not resumed for nearly two centuries, notwithstanding the ceaseless speculations by which that period was characterized. The names of those early thinkers deserve perhaps to be recorded, for even when the light of their teaching reaches us through the distance of twenty-three centuries, they shine like luminaries of the first magnitude. It was Anaxagoras who had the honor of reviving the inquiry, and who was the first to arrive by the examination of nature at the idea of an infinite *personal* intelligence as its Creator and Governor. The inquiry did not, however, end with him. Suspended during the half century of intellectual scepticism and of political commotion which followed him, it was resumed by Plato, who superadded (as we before stated) to the ideas of life and personality in God the idea of moral attributes and a moral providence.

These three ideas, the discovery of the three individuals whose names have been mentioned, form the collective conception which the period of the most acute speculation could attain in reference to the Divine Being;* and the subject affords us matter for serious thought; for you will notice that, while by dint of successive efforts they obtained an indistinct glimpse of God's attributes, they were hardly

---

* It may be thought that some notice ought to have been taken of the theology of Aristotle. The omission was made when the Sermon was written, because Aristotle's pantheism seemed a step backwards from Plato's monotheism, rather than an advance. The writer would at the present time justify the omission, because, under the view which he now takes of Plato's theology, he would consider Aristotle's view a mere variation in statement, not in the point of view; the Divinity of Plato being the highest *formal* cause, that of Aristotle the highest *final* cause. In neither view is Deity a person; in both an abstraction.

able to make any discovery of the relations which He sustains toward man. Compare with their views the language of Isaiah in our text, and you will find his views as much superior to theirs in conception as they are loftier in expression. "Thus saith the high and lofty One that inhabiteth eternity, whose name is Holy; I dwell in the high and holy place; with him also that is of a contrite and humble spirit, to revive the spirit of the humble and to revive the heart of the contrite ones." The Greeks could discover in an humble degree the One that inhabiteth eternity; in some sense they could even perceive His moral attributes and see that His name is Holy; but they could know nothing of his dwelling with the contrite. Nay, that which is more striking is the fact that, in proportion as they attained to the one idea, they lost the other. Under the vulgar polytheism they had been wont to regard the gods as conversant with the affairs of men, as rewarders of good and avengers of evil; but when they rose to the idea of God as a cause, as infinite, as intelligent, they conceived Him as reposing in the perfection of His own blessedness, and as regardless of the insignificant affairs of mortals. It was revelation alone which combined the two ideas. The very combination is one of those thoughts of God, which are not man's thoughts, to which man could not by mere reason penetrate.

2. We proceed accordingly to notice the discoveries on this subject gradually unfolded by Divine Revelation,—the successive steps of advancing knowledge through which Providence has brought mankind.

We need not pause to examine what was the information which the Hebrew patriarchs were permitted to enjoy concerning the greatness and condescension of God, because there is one remarkable epoch pointed out in the Scriptures at which human knowledge on these subjects was suddenly increased.

An exiled shepherd was the subject of the revelation.* A vision was presented of a burning bush, and to the mind of the praying shepherd, and through him to his nation and the world, was vouchsafed the knowledge of that awe-inspiring and mysterious attribute of Deity, "I am that I am." We cannot measure the amount of religious knowledge which Moses previously possessed. Undoubtedly he had some conception of the spirituality and unity of God and of the responsibility of man, and it is hardly to be supposed that his experience had not sometimes brought home to him the consciousness of the ancient patriarch, "Truly God was here, and I knew it not." But it was then for the first time that he was taught to realize distinctly that the God of Abraham was infinite and omnipresent, and that these attributes were the pledge that He would fulfil His promises to the Jewish people. The very scene of the vision was suited to its subject. It lay among the solitudes of Horeb. Moses was alone, surrounded only by those grand types of unchanging nature, "the everlasting hills." And he was made to feel that, even there, God was with him; that there was no such thing as solitude, that every spot through the expanse of space was inhabited by the Almighty; that at any moment the heavens might reveal to him His presence; that though fancying himself in loneliness, he was in contact with his Maker.

And if any one would wish to remark the effect which that revelation of the infinity of God had on the mind of Moses, let him turn to the record of the Creation which he has prefixed to his writings,—a record which contains con-

---

* Exodus 3 : 2–9.

ceptions so sublime that they have even been quoted with admiration by the heathen critic;* and which, whatever may be the interpretation which science shall ultimately put upon them, must remain the most elevated conception of Creation ever presented; for there are many noble thoughts of God in the Scriptures, and many which must excite marvel, but there is not one more noble than Moses's narrative of the dawn of Creation.† That conception is grand which the Bible affords us when it presents the thought of the Divine Being as the sustainer of the whole universe, animate and inanimate, that in Him we live, and move, and have our being; so that, whether we look out upon the stars as they march in their brightness, or hear the winds as they sweep by in their rushings, or watch the waters as they flow in their tides, all, all are sustained by His hand and regulated by His will; so that there moves not a being even on the outskirts of Creation that does not draw animation from His fulness, for He filleth all in all. That is, if possible, a still more sublime conception—the vision which the Bible shows us in the distant future of the spreading of a great white throne: upon it sits the Ancient of Days; sea and land yield up, in the twinkling of an eye, the myriads which are held in the sepulchre; all are marshalled there—the kings and great ones renowned in

---

* Longinus on the Sublime, ch. 9.

† The thought, and in some cases the language, of the next page is borrowed from one or two sermons of the Rev. H. Melvill. I wish to take this opportunity of saying that, shortly before writing the present Sermon, I had read several sermons by this eloquent preacher; and I believe that, in composing the second head of this Sermon, I borrowed a few scattered expressions from him consciously, and probably some unconsciously. I wish to make this general acknowledgment, not being able to remember the exact references to his works.

history, the noble and the mean, the learned and the ignorant. The remotest corners of the earth, the most distant ages of time, contribute their spirits to the tremendous gathering of those who are to receive their everlasting doom. Inanimate nature seems (as it were) to sympathize with the solemn scene; for the earth and the heavens flee away, and there is found no place for them.

But, grand as are these conceptions, that seems still grander where Moses makes us spectators of the birth of created nature. He calls up to our imagination a season in the distant depths of a past eternity, when the assemblage of stars and of systems which strew the fields of space did not exist; when no glorious or undying spirits, angelic or human, lived to comprehend the God that had given them being. Nothing ever broke that wondrous silence save the voice of the Eternal One, who existed from the unfathomable depths of eternity. God was there then, as now, in three Persons, the ever-blessed Father, Son, and Holy Ghost. But the universe held only God, and in that Divine Being was the attribute of benevolence, infinite then as now; and that benevolence craved the being girt round by dependent creatures. It seemed not good to God to continue alone; the sublime loneliness was infringed; the word was spoken, and the depths of space became strewed with worlds; and immortal spirits, sparklings of His infinity, thronged His presence.* "The morning stars sang together, and all the sons of God shouted for joy." Such is the conception of the Divine Being which Moses has presented to us, and from it we may understand, better

---

* The idea in this passage was suggested by some remarks in Dr. Harris's "Preadamite Earth."

than from words, what was the revelation of the Infinite vouchsafed to him at Horeb.

We may also be sure that we shall not err if we regard Moses's state of religious knowledge as the highest limit to which mankind attained under the Law. In tracing the growth of ideas on this subject in later dispensations, we should not forget that the progress of man's religious knowledge has been marked by epochs rather than by continuous advancement. Some new revelation has given a sudden expansion to it, and it has then remained stationary for a period. Thus it should be remembered that the Law is separated from the age of the Prophets, which commenced with Samuel, by a chasm of nearly four hundred years,—a period almost as long as that which intervened between the voice of the last prophet, Malachi, and the coming of St. John the Baptist,—an interval during which revelation was wholly silent, in which, in the striking language of Scripture, "there was no open vision." Accordingly, when the voice of inspiration again is heard, the ideas which it utters are really, on some subjects, so much in advance of those which were presented by the Law, that there is correctness in the ancient view which regarded the Law and the Prophets as two distinct revelations. Though this improvement in religious knowledge is especially evident in the light which the Psalms and Prophecies cast on the meaning of sacrifice and on the coming of the Messiah, yet some degree of advance may also be traced in those doctrines which form the present subject of consideration.

For no one can read the Psalms without feeling that their writers, and especially David, had more elevated views, both of the greatness of God and His condescending love, than were furnished to pious minds under the Law. What frequent intimations, for example, are to be found that the contemplation of the visible creation ought to have an efficacy in inspiring belief in God, and adoration and love of His perfect attributes; yet how equally marked is David's unwavering faith in the presence of God with the heart of the praying worshipper! His own early life had probably prepared his mind for inspirations of this character. We can easily imagine that when wandering together with a band of attendants amid the rugged highlands that mark the physical features of Southern Palestine, hunted down by the malice of an implacable foe, he would be compelled to pass many days and nights with no other employment than the communings with nature, and with the God of nature through his works, which would arise in his poetic and pious mind. It was in deserts, indeed, similar to those in which he was an exile, that the ancestors of his nation had wandered for a generation; yet they had recorded no conceptions such as his. Their faith in God's protection may have been as real as his, but it cannot have been as elevated; for they believed in a God whose visible manifestation, as a cloud by day and a meteor by night, was the proof of his presence; but David, without such miraculous evidence, was able to rise, by meditation on His works, to the conviction of the immeasurable greatness of that Being, of whose support he felt nevertheless sure, and whose presence, in some real but humble sense, he found within his own heart.

What can exceed, for example, his conception of God expressed in the words, "The heavens declare the glory of God, and the firmament showeth His handiwork. Day unto day uttereth speech, and night unto night showeth knowledge. Their line is gone out through all the earth, and their words to the end of the world."* And, passing

* Psalm 19 : 1–4.

by means of analogy from these physical to the moral laws of God's judgments, he adds, "More to be desired are they than gold, yea, than much fine gold; by them is thy servant taught, and in keeping of them there is great reward." Again: "When I consider the heavens, the works of Thy hands, the moon and the stars which Thou hast ordained, what is man that Thou art mindful of him, or the son of man, that Thou so regardest him? Thou hast made him a little lower than the angels, and hast crowned him with glory and honor."* Again: "Who is so great a God as our God, that hath His dwelling so high, and yet humbleth Himself to behold the things there are in heaven and on earth?"† How noble are these thoughts concerning God's greatness! Similarly concerning His eternity, in the passages: "From everlasting to everlasting Thou art God." "Thy kingdom is an everlasting kingdom, and Thy dominion endureth throughout all generations. The Lord upholdeth all that fall, and raiseth up all those that be bowed down."‡ And (to allude only to one more instance) what words could more nobly express the overwhelming contemplation of God's omnipresence, or more separate the infinite Being from the finite, than the passage, "Whither shall I go from Thy spirit, or whither shall I flee from Thy presence? If I ascend up into heaven, Thou art there; if I make my bed in hell, behold Thou art there. If I take the wings of the morning, and dwell in the uttermost parts of the sea, even there shall Thy hand lead me, and Thy right hand shall hold me. If I say, Surely the darkness shall cover me; even the night shall be light about me; yea, the darkness hideth

---

* Psalm 8 : 3–5.
† Ps. 113 : 5, 6 (Prayer-book, *i. e.* Miles Coverdale's Version).
‡ Psalm 90 : 2; 145 : 14.

not from Thee; but the night shineth as the day; the darkness and light are both alike to Thee."*

This is the utterance surely of one who felt to the fullest that if he were endowed with unlimited powers of motion, he could never for a lonely instant escape from God; that God would remain at the spot which he had left, and be found at the place which he had reached, that the darkness of the midnight shrouded not from Him, that the depths of the heart lay open to His inspection; that no act could escape his observation, no wickedness be so stealthy as to go undetected by Him.

If we pass from the Psalms to the Prophets, from David to Isaiah, though we do not find much advance in the conception of the greatness of God, yet there is a growing clearness in the ideas concerning God's holiness and His condescension in dwelling with the sinful. "The Holy One of Israel" is (if the expression may be allowed) Isaiah's favorite mode of speaking of the Almighty. We may perhaps trace the cause of this peculiar mode of thought (as we did previously in the case of Moses) to the vision which was given to him at his call to the prophetic office.† The Lord appeared to him on His throne; and seraphim stood around crying, "Holy, Holy, Holy." We cannot wonder that the prophet stood confounded, and said, "Woe is me, for I am a man of unclean lips." It was the manifestation of the holiness of God which convinced Isaiah of his own imperfection. And the seraph took a live coal, and touched the lips whose uncleanness he had bewailed, and pronounced that his sin was purged. Can we wonder that

---

* Psalm 139 : 8–12. David's view of nature must be regarded as an artistic or emotional, rather than as a scientific one.

† Isaiah 6.

such a sight as this,—the radiant form of the Lord, throned in fire and cloud, with angels chanting their song of triumph,—should leave a lasting impression on the prophet, alike of God's holiness and of human sinfulness; of God's majesty and condescension? or need we seek for any other epoch than this, when the revelation was made to him, which he has expressed in our text, that "the high and holy One who inhabiteth eternity, dwells with the humble, and revives the spirit of the contrite?"

If we leave the early dispensations, and pass on to the Christian, we shall find that the thoughts of men are again widened. Christianity indeed adds little to our idea of the infinity of God; but it adds much to the idea of His mercy and condescension. What a palpable proof, for example, of God's love is seen in the fact of the appointment of a human mediator! If Christ had been merely Divine, if He were unallied with ourselves, if He had never taken our nature nor experienced our trials, could we have confidence in committing ourselves to Him? If you would encourage me to carry my sorrows to such a mediator, you could only point to His infinite greatness and amazing power; but this would merely increase my misgiving whether He would condescend to notice such an unworthy being as myself; for in proportion as you raise my conception of Him, you remove Him from all companionship with the sinful. But when I see the Word made flesh, it is a pledge that whatever is human must come within the sphere of His mercy. Watch those thirty-three years of His earthly life; see Him tasting deep of every sorrow; sustaining every human relation; bearing others' sufferings, and carrying their trials, never refusing mercy to the vilest, healing the most impure; and remember that the same being, though gone on high, is Jesus still; so that now, within the very shrine of the eternal glory, there dwells one in human form, with all the strength of human sympathies, and the remembrance of human trials, interceding for us; infinite in power, because God, unceasing in sympathy, because man: and tell me if you do not feel confidence to approach the mercy-seat, and cast your load of sin before Him; tell me if you do not now understand, with a fulness which even Moses and David and Isaiah could never realize, that He who inhabiteth eternity indeed dwells with the humble, and is willing to revive the heart of the contrite.

How marvellously also are we made to feel the condescension of God, by the doctrine that Christ has given another Comforter to dwell in the human heart! For though His miraculous gifts are no longer the sensible proof of the Spirit's presence, yet his moral influences are as real now as ever. We are not awed by unearthly spectacles, nor convinced by supernatural evidence; but wherever there is a heart touched with a sense of its own sinfulness, or longing with anxious earnestness to be delivered from the slavery of sin, or conscious that it has, indeed, been turned from the love of sin to the love of God, there the Spirit's operation is manifested; there His influences are found, even now. Ask any one who is the blessed recipient of those deep searchings of hearts, and he will respond in the words of the Prophet: "Truly, the high and lofty One that inhabiteth eternity, whose name is holy, dwelleth with him that is of a humble and contrite spirit."

3. We have now completed our survey of the information which is supplied to us in reference to the Divine greatness and condescension by Reason and Revelation. There remains, however, another medium of communication. Startling as it may seem, we can show that the discoveries of modern Science have opened views of the Divine greatness

which even add something to that which Divine revelation itself supplies. In some respects, indeed, the discoveries of science fall immeasurably below those of revelation, but in other subjects not so. And it is a new, and I hope to make it appear an instructive, view of science, to regard it as a *revelation*, differing only from the Divine one in being communicated without supernatural inspiration, through the agency of human genius. In truth, the discoveries which the human mind makes cannot be regarded as an accident. Though they are not direct gifts from the Almighty, yet a believer in a moral governor must admit them to be part of the scheme of Providence. If discoveries merely related to the material welfare of man, we could conceive it barely possible that they arose by accident; but when we see that they bear directly on human civilization, and on the advancement of that which is eternal in man, and seem marvellously adapted to promote such advancement, we might establish an argument from final causes in favor of their occurrence by design. Thus, for example, when we look back on an event like the Protestant Reformation, we cannot think that the Almighty was indifferent to an act or series of acts which must ever stand out as the most glorious of all revolutions, the charter of the intellectual, social, and religious freedom of entire nations; and when we see how pre-eminently the single invention of the art of printing has contributed to diffuse that freedom, and to establish it as the invaluable inheritance of man, can the mind that believes in a God of Providence doubt that such coincidences are in the hands of Him in whom we live, and move, and have our being? Nor will such an individual read a Divine purpose in those discoveries only which contribute to utility; he will perceive it in those also which have no other use than merely to enlarge the range of human thought. Accordingly, in this view, Science becomes a kind of revelation from God, given by natural means, yet ordered of Providence.

Thus, for example,* what an advance has been made beyond even the knowledge which the Scripture writers possessed, in our conception of the infinite greatness of God, by the discoveries of modern astronomy! If one of the ancient prophets, who conceived this world to be the principal body in the universe, and who had no satisfactory view either of the sky or the bodies which moved in it, could have been taught those truths which observations by the telescope and the calculations of modern analysis have been instrumental in establishing, how immeasurably would his conceptions of the Divine Being have been enlarged, and his reverence for His greatness have been enhanced! Give him to understand that the earth is but an inferior member in a small system of stars; unfold to him the *plan* of that system of which, viewing it from the earth, he only sees the *section;* then show him that larger system in which our whole solar system forms a mere speck; carry on his thoughts still farther to systems situated at such a distance that the glittering millions of the bodies which compose them are undistinguishable except as a spot of nebulous light; place before him the extraordinary fact that the ray of light, travelling at a velocity which baffles the powers of imagination, though not of calculation, probably must have left those bodies even thousands of years ago; convince him further that the chief part of these statements are not theory, but really matters of measurement, depending originally upon facts the most obvious, and computations the most simple, and would not he be

* Compare Hitchcock's Geology and Religion (Lect. 13).

prepared to admit that God has, indeed, unfolded to these later ages of the world conceptions of His own infinity and majesty, which prophets of old waited for and sought, but never found?

Nor is it only by means of the infinite in greatness, but also by the infinite in smallness, that we learn the nature of God; for as the telescope has revealed to us the one, so the microscope has laid bare the other. Each has enlarged human power so as to confer on it (we might almost say) a new sense. When we find a world of minute life discovered to us, unperceived by our unassisted senses,—nay, when we find that with every successive increase in the power of the instrument, a world of still more and more minute life is laid bare, so as to seem to have no limit to its immeasurable minuteness, just as in the other direction the series of worlds seems to reach to infinity in their immeasurable remoteness; when we find these minute beings wondrous in structure, and surrounded abundantly by all that is adapted to their wants, we see that the vastness of God's care reaches to the fleeting and insignificant unit of an insignificant race; the very atom is not overlooked; and we begin to understand in a deeper sense what is meant by the declaration that God is "the high and lofty One, that inhabiteth eternity."

If we pass from these sciences, others meet us which teach the same doctrine. For as Astronomy has stretched our conceptions to apprehend the extent of the physical universe in *space*, so does Geology expand them in relation to our ideas of *time*. It used to be considered that, about six thousand years ago, the earth and the whole material universe were spoken into existence in a moment of time. We now understand that the Scripture account, which was supposed to imply this, can only relate to the preparation of this earth for the habitation of man, not to its original creation. For science has proved, by irrefragable evidence,* that the first act of creation must be referred to a period indefinitely but immensely remote; and that successive ages have passed over this globe, during which it has been the seat of numerous systems of organic life, differing from one another, yet all linked into a great system by a most perfect unity. The revolution of thought, which reduced the world to its true position in the universe of space, did not more immeasurably enlarge our ideas of the Divine Being than this has, which has reduced the era of human history to its true position in the immensity of time.

And as Astronomy has revealed to us the infinity of the *present* creation, and Geology the vastness of the *past*, so has the science of Mathematics laid open to our view the infinite wisdom which has provided for the *future*. If any branch of knowledge appeared eminently unlikely to unfold to us any information about God, you would think it would be that system of symbolic formulæ and abstract notions, which seems to stand in utter isolation alike from nature and from man. And yet when we apply it to predict the attractions of the heavenly bodies in periods yet to come, it unfolds to us some results of extraordinary grandeur. When we trace the effects of the mutual disturbances of the planets, we seem to approach a mighty catastrophe, which their mutual action will at some time bring about; yet as we pursue our calculations we arrive at a few unassuming formulæ, which, interpreted by reason, reveal to us the infinite wisdom of God; for we find that not only vast cycles of time are established in which these disturbances, when verging on the catastrophe which we dread, shall

---

* See a note appended to Sermon III of this volume.

begin to be reversed in their effect, and thus restore the whole solar system to the position which it originally occupied; but also that, by an exquisitely-contrived plan of compensation, the stability of all those elements which are essential to the safety of the system is permanently guaranteed.* Who can contemplate these amazing results, which manifest the infinite contrivance of the Almighty Architect, without a feeling of devout thankfulness that we have been permitted thus to discover traces of the high and lofty One who inhabiteth eternity!

The illustrations which have hitherto been given show how Physical Science has revealed the *infinity* of God; we might add, also, that Mental Science has equally revealed His attribute of *holiness*. For as we study the microcosm, man, we remark the existence of a faculty there, acting as God's vicegerent, imperatively forbidding sin; and as we study man in the larger relations of society, we observe that a system of pleasure and pains has been annexed to virtue and vice, of such a character that virtue is made its own reward, and vice its own punishment.† So that if we before learned to understand that God was high and lofty, and that He inhabiteth eternity, we now are taught that " His name is holy."

These remarks must suffice in proof of the assertion that Science has opened to us some views of the Divine Being

---

* The allusion is to Lagrange's theorems on the stability of the inclinations of the planetary orbits, the conservation of the mean distances and periods, and the stability of the eccentricities. They are stated in most works on the planetary theory; *e. g.* in Airy's Mathematical Tracts, or Pratt's Mechanical Philosophy. Sir J. Herschel has attempted to give a popular explanation of them in his work on Astronomy (part ii. ch. 12, 13).

† Compare Chalmers's Bridgewater Treatise (ch. 1–3).

which even surpass those which are furnished by Scripture itself. It will, however, moderate the pride with which we might be in danger of regarding Science, if we remark the deficiencies of it as a revelation in certain other respects. It reveals to us Him who inhabiteth eternity, but it tells us nothing of His willingness to dwell with the humble; it reveals to us general laws, it cannot teach a special Providence; it may show us man's misery and his need of penitence, but it is the Bible alone which can tell us that the Infinite is approachable by prayer, and willing to revive the heart of the contrite.

We have now traced through the world's history the successive discoveries which Reason, Revelation, and Science have made concerning the greatness and condescension of God. It only remains, in conclusion, to draw from the subject very briefly some practical instruction on the motives and means of religious living, which may tend to advance us a step in the path toward that world where the infinite God shall no longer be comprehended imperfectly by laborious processes of inference, but in the simple power of an undimmed intuition shall be known face to face, and be seen as He is.

If it be indeed a fact that the high and lofty One that inhabiteth eternity is willing to dwell with man, what appeals it ought to make both to our fears and our hopes! The thought ought to warn us that nothing can escape His observation, that all our sins lie open to His searching gaze, and are registered in the book of His undying memory; and if He be represented to us as caring so much for man that He has set His heart upon him, what can we expect if we slight the salvation which He proffers us but the vengeance of His overwhelming greatness?

On the other hand, what a blessed hope it furnishes to

each one of us! The infinite God is willing to become our friend, and to make our hearts His dwelling-place. What privilege can be so exalted as this? We see some men moved by an honorable ambition covet the prize of worldly praise, or the friendship of the great of the earth; and others withdraw themselves from the hurry of society to commune with the minds of former generations, as they meditate on the works which hand down to us their thoughts. But how immeasurably nobler is the privilege of entering into communion with Him that inhabiteth eternity! how much more exalted an employment to retire to seek of Him to send his Spirit to dwell within our hearts!

The method whereby this privilege is open to us is plain. He dwelleth with him that is of a contrite and humble spirit. The one requisite for its attainment is that we really feel our own sinfulness, and ask for His presence. Therefore, if there be one of us more conscious than his fellows of his own exceeding great needs, and well-nigh desponding that so high a privilege can be for him, he may take comfort that the Almighty does not seek for worthiness in us,—He only asks for willingness; that He will be found of all who approach Him in earnest penitent prayer, for He will revive the spirit of the humble, and the heart of the contrite ones. How great a means also of realizing His nearness should we find it, if we were to bear before us the thought of His greatness and omnipresence! What a reality would it throw into our prayers, either in public worship or in private, if we forced upon ourselves, as we bow the knee in supplication, the thought of Him whose presence we are invoking! How would it hallow our lives if we possessed within us, amid the pressure of business or the whirl of fashion, the vision of the infinite God; if we remembered that no fretting cares, no innocent excitement, need shut us out from His presence; nay, that from amid the hurry of the multitude and the tumults of life there is a hearing for every humble heart in the heavenly temple; that the unuttered breathings of the most secret wants of every contrite spirit are seen, and known, and heard, and answered afar off, in that place where the Babel tumult of earth is hushed, and the stillness of the sacred presence is unbroken save by the seraph chant of "Holy, Holy, Holy," or by the chorus of reverent praises, which rises from the ten thousand times ten thousand of the redeemed spirits of the just made perfect!

It is indeed a joyful thought that God so inhabiteth eternity, that travel where I may in unlimited space, I can never reach the lonely spot where He is not present as my guardian, never find the solitary scene where He is not as watchful over me as if the universe were a void, and myself its sole inhabitant; and, therefore, I know that though I may live among the humblest, I am as much observed of Him as a monarch on his throne; that when I go to my daily toil, or say my daily prayer, when I lie down or rise up, I am cared for of Him; so that I cannot weep the tear which He sees not, nor feel the pang which he notes not, nor breath the prayer which He hears not.

NOTE,

On the Theological Ideas of some early Greek Philosophers.

It has been stated in a note to the preceding Sermon, that the writer of it has been subsequently led to alter his views in reference to the tenets which some of the old Greek philosophers, named in it, entertained on the subject of Deity. He thinks that he had attributed to them more than they consciously knew. He had made them think on theological subjects in too modern a spirit. It is important to remember that the study of those early thinkers is like the study of a fossil world. The same words which we now use were used by them, but with wholly different meanings. The difficulty is really not to find dissimilarities between them and ourselves, but rather to find points of agreement.

The points, accordingly, in which the preceding Sermon appears to exceed the limits of fact are:—

1st. In attributing to Thales, on the strength of Plato's remark in the "Philebus," *conscious* attempts to speculate on theology; whereas his speculations partook rather merely of the character of ontology or cosmogony, and can only be regarded at relating to theology in the single point where he touched on the idea of *power* or efficient cause, and identified it with the material cause, making both reside in Water, thus attributing a kind of soul to Nature.

2d. In making the Νοῦς of Anaxagoras's system to be a *personal* intelligence, whereas it was probably hardly more than the idea of *order* or *law*, presiding over nature, in contradistinction to Heracleitus's view of constant flux in phenomena.

3d. In making Plato to have regarded Deity as a *person*, and interpreting the term ἀγαθός in his description to refer to *moral* qualities. His God was rather the mere principle of Divinity, and the *goodness* of his Deity was only *order* or *harmony*. The supreme ἰδέα seems to have been at once the supreme type of goodness and the supreme formal cause, GOD. So that the 'Ιδέα, arising first in Plato's mind as a mode of accounting for reminiscence (as in the "Meno"); then becoming a mode in controversy for refuting scepticism (as in the Theætetus); next, regarded as having a real existence in nature and in knowledge (as in the "Republic"), analogous to our "law of nature" or our "hypothesis;" lastly (in the "Philebus"), came to be regarded as the supreme cause, the most abstract harmony, the GOD. The whole account of the Deity in the "Republic" (b. ii) is explicable on this hypothesis.

Thus all philosophical theology in Greece was Pantheistic, *i. e.* if Pantheism be made to mean any theory which admits an *impersonal* first cause, and to include, (1st.) The theory which teaches an *anima mundi;* (2d.) That which regards God as the sum total of all that exists (Pantheism proper); and (3d.) That which regards Deity as an abstraction, synonymous with the idea of perfection. Thales might possibly represent the first of these views; the Eleatics the second; Anaxagoras, Plato, and Aristotle, the third.

(pp. 34–42.) It will be observed that no reference is made in this Sermon to the modern German discussion concerning the authorship, integrity, and date of the books of the Hebrew Canon here quoted. Such criticisms were partly unknown to the writer at the time when the Sermon was composed (1855); but they do not so much affect the question here discussed concerning the conceptions to which the Jews attained about God as the respective dates at which those conceptions became known. The view taken in the Sermon, being the one prevalent in England, supposes the general integrity of the Masoretic Canon, and assumes also that the different conceptions of Deity, commonly called the Elohistic and Jehovistic, related more to difference of thought than of time, not being so much restricted to particular epochs of Jewish literature, but rather marking respectively different aspects of belief, the priestly and the prophetic, the ethnic and the revealed. (The reference, however, in p. 41, of the authorship of the 139th Psalm to David on the strength of its title is most probably erroneous, on account of the Aramaisms which it contains.) This interesting subject of the Jewish names for Deity may be studied in Hengstenberg's "Authentie des Pentateuches" (vol. i); De Wette's "Introduction to the Old Testament" (Eng. Trans., Part III, B. 1, ch. 1); Keil's "Lehrbuch" (p. 82); and Donaldson's "Christian Orthodoxy" (Appendix III).

SERMON II.

DIVINE PROVIDENCE IN GENERAL LAWS.

(PREACHED BEFORE THE UNIVERSITY, DECEMBER 13, 1857.)

ACTS 17:28.

"In Him we live, and move, and have our being."

ST. PAUL'S visit to Athens, on which occasion the speech was uttered of which these words form a part, is one of the most interesting passages of apostolic adventure which Scripture history has preserved to us. If St. Luke had merely informed us that the Apostle, in the course of his missionary travels, visited the two great centres of ancient power and civilization, Athens and Rome, but had omitted to record his acts and speeches under those circumstances, we should have been probably more disappointed by the omission of such a narrative than by that of any other portion of his eventful life. We should have wished to know how St. Paul felt on those occasions as a man, and how he acted as an Apostle. We could even be content to have lost the narrative of his visit to Rome, rather than that of his speech at Athens; for when we had heard that Rome was visited by him when a prisoner, we might be certain that this circumstance would interfere with his freedom of action, and with the expression of his missionary sympathy. But we should have earnestly coveted to know what the Apostle did and said in bringing Christianity for the first time into contact with the religion and philosophy of Greece. And, therefore, we must set a special value upon the precise narrative which St. Luke has left us, which, in thrilling interest, equals all that we could have anticipated.

The Apostle visited Athens in circumstances which (as we have already hinted) were wholly unlike those which characterized his visit to Rome. He entered Athens as a freeman, happy in the joyful recollection of escape from recent perils in Northern Greece. He was unattended by companions. He wandered alone through the city, undisturbed in his examination of it. And then, after his day's wanderings, he is described as betaking himself to the Agora, and breaking out into a public address to the gathered crowd. It is Paul the traveller no longer; it is Paul the Apostle. The thoughts, struggling like a pent-up fire, now express themselves in words. It is Paul trying to do his Lord's work in the centre of the world's civilization. It is a Christian Apostle, and that Apostle Paul, bringing Christianity into contact with Greek religion and Greek philosophy, in a city hallowed by historical associations, a sanctuary of art, a centre, even in its fall, of intellectual glory.

No scene can be more interesting. Yet we are apt to allow our recollections of an earlier age of Athenian history to interfere with a vivid realization of the scene and the audience which presented themselves to St. Paul. We are accustomed to think of Athens only as she was in the zenith of her power; when, flushed with the victory of freedom, and guided by the consummate ability of her statesmen, she raised herself to the head of a vast colonial empire; when her foreign power was but the parallel to that more enduring intellectual empire which existed at home,—the empire of art, of thought, of liberty. But when St. Paul visited Athens, nearly five hundred years had passed since the noonday of Athenian glory. Her empire had been destroyed; her commerce had disappeared; her harbors were already beginning to be choked with sand; and her territory had been absorbed in that of the power which had converted the Mediterranean into a Roman lake, from which, as from a centre, succeeding waves of conquest were overflowing the earth. The rays of intellectual glory still, however, lingered round the setting splendor of Athens; and the young nobility of Rome came to reside there as in a sort of University. It was with some of these persons, men of speculative habits of thought, that St. Paul was now brought into contact. His audience consisted not only of the persons who in the evening might be reposing under the plane-trees of the Agora, but embraced also certain students of the Stoic and Epicurean creeds. St. Paul knew this, and we may observe a wonderful adaptation to the tastes or the errors of each sect in the discourse of which his historian has presented the substance.* The Apostle was taken up to Mars' Hill; and, standing on those rock-hewn steps which

---

* Some critics have supposed that St. Luke has here put into St. Paul's mouth a speech such as he was likely to have delivered, according to the dramatic method so common in the historians of that time. The supposition has arisen from the improbability that the logical mind of the Apostle Paul would be susceptible to the influences of poetry, and scenery, and other circumstances, of which the speech exhibits traces. There seems, however, equal internal probability against such a supposition.

formed then, as now, the only approach to the hill, he spoke with somewhat perhaps of the air of dignity which the great master has expressed in his cartoon.* The scene was one of the most striking, the most poetical in history. His very opening words seemed to convey the impression which the sight of the city had made upon his mind, and with consummate tact he seized on a familiar illustration. "Men of Athens," he said, "all things which I behold bear witness to your carefulness in religion.† For as I was passing through your city, and beholding the objects of your worship, I found amongst them an altar to the unknown God. This is the God whom I declare unto you."

It is surely no idle fancy to suppose that the words which follow may suggest to us the belief that the eye of the Apostle at that moment glanced from the eminence on which he stood to the citadel which rose in queen-like stateliness before him, covered with those many temples sparkling at that time in crystalline whiteness, the remains of which still attract the traveller to gaze on their majestic outlines, beautiful even in their ruins; and from that exhibition of industry and wealth, devoted to the service of the Greek religion, the Apostle's eye may have glanced away

* It is not implied by these words that the Apostle's figure, marred by his "thorn in the flesh," can have been so noble as the ideal which Raffaelle (copying Masaccio) has presented; but it may be reasonably presumed that his *moral* dignity and bearing have not been inaptly depicted in that master's cartoon.

† It is generally known that the words used by St. Paul (ὡς δεισιδαιμονεστέρους) were intended as a compliment, and not as a reproach, as the English translators have made them mean, in mistranslating them by the words "too superstitious." The paraphrase of St. Paul's speech here given is borrowed from the work on St. Paul's life and epistles by Conybeare and Howson (i, 401).

again, as if by contrast, to the works of Nature; to the beautiful plain, bounded by its marble mountains on the one side, and to the sea and distant coast, which formed the girdle of the horizon, on the other. And it may have been under the impression of these feelings that the Apostle continued: "God, who made the world, and all things in it, seeing that He is Lord of heaven and earth, dwelleth not in temples made with hands" (such as then were rising in majesty before him); "neither is He served by the hands of men, as though he needed anything, for it is He that giveth unto all life and breath, and all things. And He made of one blood all the nations of mankind to dwell upon the face of the whole earth, and ordained to each the appointed seasons of their existence, and the bounds of their habitation; that they should seek God, if haply they might feel after Him, and find Him, though He be not far from every one of us: for in Him we live, and move, and have our being; as certain of your own poets have said, 'For we are also His offspring.' If, then, we are the offspring of God, we ought not to think that the Godhead is like unto gold, or silver, or stone, graven by the art and device of man." And then the Apostle proceeded to preach to that frivolous crowd, with intensest earnestness, a judgment to come, and the future life.

We need pursue the subject no further. The narrative must have an enduring interest, not only as one of the few discourses in which an Apostle has shown a susceptibility to the influences of scenery, but still more as a specimen of the mode in which Christianity was presented for reception to a heathen, an educated, a philosophical audience.

Those two schools of Stoics and Epicureans, before whom St. Paul spoke, divided at that period the sympathies and the belief of thinking men. Speculation into the deeper mysteries of existence and the problem of the universe had long since disappeared.* Experience had convinced mankind of the futility of those attempts which speculative philosophy had made to solve such problems; and the attempt had been laid aside, to be again and again resumed in after ages, with the same ill success. Men had accordingly begun to feel that the business of man was not to speculate, but to act: and so they were divided into two classes; the one of which† comprised those who bore lightly the sight of woe, and passed life in an elegant selfishness; the other,‡ those who, with haughty self-respect, yet in the main with a serious view of life, acted upon convictions of duty, of the origin and of the future reward of which they were ignorant. Yet whatever may have been the excellence of either system, both alike were godless; both were

---

* The Epicurean and Stoic philosophies included indeed speculations into Physics, as well as Logic and Ethics; but the inquiries were not only subordinate, but were conducted without the ontological speculations which had belonged to the philosophies of the Eleatics, of Plato, and of Aristotle. It is to the Stoics that we mainly owe the psychological character which has always since marked ethical investigations. The Stoic ethics are in tone modern; all previous philosophies are like the fossil remains of an extinct creation. The best account of the Stoic school is to be found in an essay by Sir Alexander Grant, in the Oxford Essays for 1858. It is probably to the influence of the Stoic philosophy, impressed on Greek thought and on the Greek language, which St. Paul consciously or unconsciously received in his youth at the schools of Tarsus, that we must attribute the internal, the psychological aspect of sin, which that Apostle presents in the 7th of Romans and elsewhere. Sin is not, in his view, merely transgression against Heaven; it is internal, moral disorganization. Such a view does not lower the Apostle's inspiration. The anointing Spirit did not obliterate peculiarities of knowledge or of mind in the individuals whom He inspired, but condescended to consecrate their various gifts.

† The Epicureans.   ‡ The Stoics.

schemes of life which shut out the world invisible, and were constructed in disbelief of a personal Providence. To each of these classes St. Paul addressed his discourse. In the lofty and serene Deity, who disdained to dwell in the earthly temple, the Epicurean would find echoed his own belief; and in the idea of an orderly system, the Stoic would recognize his own hypothesis of Fate. But St. Paul led them beyond these notions. He at once demolished the Atomic theory of the one, by teaching that God was the sustaining power in nature; and the Pantheistic fatalism of the other, by the idea of the providence of a personal God. He taught to both that God was very nigh, that He was observant of every man, and would exact a personal account; that it was their privilege to feel after and find Him; that their dark and troubled spirits might break into the light of His reconciled countenance.

It is to the theory of Providence that I wish now to direct your attention. The disbelief in a personal Providence of those two old Greek schools is not peculiar to their time. The atheism of the Epicurean, the fatalism of the Stoic, have often been reproduced in the history of thought in succeeding ages of the world. Each period has its own discoveries; each great thinker strikes out his own line of investigation. And new discoveries or new lines of thought have to be adjusted with existing belief, or existing belief surrendered to them. Hence it is that, in several ages of Christian history, the progress of investigation has tinged men's views on Providence. Occasionally it has been some metaphysician, who, gazing into the mysteries of existence, and striving with his limited faculties to transcend the heights and depths of the infinite Mind, has figured to himself the universe as one magnificent whole of causes and effects, moving on from everlasting to everlasting by laws once impressed upon it, with a God perhaps to create, but no God to superintend.* Or else it has been one who has imagined to himself the universe as one infinite substance, ever in a state of evolution and development; an ocean on whose bosom phenomena arise, like so many bubbles which appear but for a moment; a volcano ejecting its contents into the air, only to receive them back again into the unknown depths of its own capacious crater.

Nor is it merely from speculations into Mental phenomena that an unchristian view of Providence has sometimes been developed; but the same thing has resulted from those real discoveries which have been made by the sciences which investigate Matter and Nature.† Such a view is far more difficult of refutation, and is more captivating than that to which allusion has just been made, in proportion as it is founded on truths which are indisputable, because admitting of verification. For among the great and marvellous discoveries which Science has made is the fact, that each region of phenomena seems to be directed by an

---

* The preceding sentence would describe the Deistical view of the last century in England and France. That which immediately follows alludes to Spinoza, and also in part to the modern schools of Schelling and Hegel. Spinoza reached Pantheism from the investigation of the infinite in *space*, while Schelling and Hegel rather arrived at from the infinite in *time*. His God was in a sabbath of perpetual rest; theirs in progress and development. Accordingly, the words "infinite substance" in the text, together with the two illustrations of "the ocean" and "the volcano," describe Spinoza's view; the words "evolution and development" apply rather to that of the two philosophers who have just been compared with him.

† The reference here intended is partly to Comte and the Positivists, and partly to the Deistical thinkers of the last century, so far as their views originated in *physical* science, as distinguishable from the *metaphysical* source of them to which allusion has already been made.

invariable law of antecedent and consequent,—cause and effect. The astronomer is able to show that one law of the most simple character governs every movement of the planetary bodies. The most subtle disturbances, as well as the most gigantic movements, are alike explicable by it. The power to unravel the history of the past, the ability to predict the future, are proofs that the magnificent system of universal law moves on from age to age unaltered. The student who directs his attention to the past history of the world, and who employs a curious and minute examination of the rocks of the globe, to read therein the history of those orders of existence which, during cycles of time of incalculable amount, have successively occupied this planet, finds that in the distant depths of its primeval history the same laws were in operation as now; matter crystallized in the same forms, heat and cold obeyed the same conditions as at present; vegetable and animal life conformed to the same structure. The mother fern then, as now, sheltered the infant leaflet, wrapped up within the coils of its own form. The animal races were created, lived, and passed away just as now. And if we pass from these studies of other planets, and of the ancient history of our own earth, to that small portion of its existence over which human history extends, the man of science here again thinks that he can exhibit laws in operation even over man. The same laws govern human society, and move in popular commotions, and find play in human motives now which acted of old. History can offer her generalizations. Man, like matter, comes under the domain of law.

Can we wonder, then, that from the earliest discovery, and with each succeeding confirmation of the permanence of Nature's laws, there has grown up a difficulty in believing that system of special Providence which the Bible unfolds?

The contrast is felt, that Science teaches general laws, the Bible, special adaptations. Science shows the undeviating character of Nature's methods: the Bible, their constant alteration in obedience to human prayers. Science is sceptical of a providential interference on the part of the Divine Being: the Bible teaches that He is not far from any one of us. "For in Him we live, and move, and have our being."

We believe that there is truth in both of these views. There is a scientific side of the theory of Providence, and a biblical view of it. The one teaches it to us as known to the natural senses; the other penetrates the darkness which hides the spiritual and the invisible;—the one tells us of God's works and government; the other, of His will and purposes. We shall be compelled to restrict our attention in the present discourse to the former, omitting the consideration of the Scripture view of Providence and its harmony with the scientific.*

I have already, in introducing the subject, sketched some proofs which science has to offer, to show that the system of nature is administered on a general plan. But the great evidence of that generality which I wish now to bring before you, arises from the circumstance that it seems plain not only that the Divine Being governs the world by general laws, but that when the violation or clashing of these general laws bears hard on individuals, He mysteriously on some occasions allows them to take their course in spite of the partial suffering which they produce. Such an illustration, while it furnishes proof of our princi-

---

* This Sermon was to have been followed by one on Special Providence, which the writer had not the opportunity of preaching. The line of inquiry which he intended to adopt is indicated in a note at the conclusion of the present Sermon.

ple, will offer also an opportunity for showing the wisdom and benevolence of such an arrangement; and thus of harmonizing the ideas of God's love and wisdom with that of His power.

We may draw some instances of the mysterious fact which we assert, from common events of Nature, such as accidents, pestilence, and the like. It will be desirable to picture to ourselves one of such scenes, in order to realize vividly the idea which we are striving to grasp. It shall be selected from one of those natural evils which fall upon man without his own fault and which he is powerless to resist. There is one of this kind recorded in history, which will always have a prominent interest, as having first awakened the religious speculations of the philosopher Goethe,* and aroused a controversy on Providence between those two gifted men whose cenotaphs lie beneath the noble dome which grateful France erected to the great of her sons.† It is that great calamity which about a century ago overwhelmed the capital of Portugal.‡

A fine autumn morning shone on the devoted city, and showed the groves and buildings, spreading up the heights, sparkling in beauty. The multitudes of its population had assembled in the churches to hear the morning mass, when suddenly an unaccustomed sound was heard, a long mysterious rumble, which grew louder as it approached; and when it seemed at hand, the whole city rocked like a ship heaving in a storm; the houses crumbled into heaps; the

---

* Lewes' "Life of Goethe," i, 31.

† "Aux grands hommes la Patrie reconnaissante," was the inscription on the frieze of the Pantheon.

‡ The earthquake occurred on All Saints' Day, 1755. The authority for the following account is Davy's "Letters on Literature." I have searched in vain in the Portuguese literature for the official statistics.

churches fell, and interred in their ruins the assembled congregations. A few escaped into the streets; but another shock speedily followed and destroyed many of them under the falling ruins. A large number fled to the edge of the sea, and took refuge on the pier; but lo! to their horror, the great earthquake wave,* travelling according to well understood principles at a slower rate than the undulation in the solid ground, rolled into the shore,—one huge wave of water many fathoms in depth. In one instant a mass of several thousand human beings was swept from that pier into the sea; and when the survivors, after the event, looked round on the scene of the catastrophe, they beheld the glorious city which but a few moments previously had been bright with beauty and life, a mass of ruins, with more than sixty thousand of its population buried in its fall. "Great and marvellous are Thy works, Lord God Almighty; who shall not fear Thee, O Lord, and glorify Thy name?"

Who is there that does not ponder on the mystery of that horror? Who does not marvel why the woe fell on that city? If you had asked mankind of old the explanation of that catastrophe, they would have asserted that it was an immediate judgment from heaven sent to overtake the guilty city, just as the barbarians of Melita judged, when, seeing the viper fasten on St. Paul's hand, they looked upon it as the messenger of heaven sent to slay the murderer who had, indeed, escaped the shipwreck, but whom vengeance suffered not to live. But our blessed Lord, once and for ever, forbade such cruel surmises concerning others, when, in alluding to a recent accident of

---

* See Daubeny's "Volcanos," part 2, ch. 32, 33; Mallet on the Dynamics of Earthquakes, in "Trans. Roy. Irish Acad. 1845."

his own time in Jerusalem, He said,* "Those eighteen on whom the tower in Siloam fell and slew them, think ye they were sinners above all men that dwelt in Jerusalem? I tell you nay." The lesson was not for them to judge others, but to take warning to be ready themselves.

We may well believe, indeed, that our adorable Saviour enunciated in this passage, and in another similar one, which relates to the man who was born blind,† a great and mysterious truth, which, like so many other great truths of the Bible, has been marvellously corroborated by the discoveries of modern science. That truth is, that not all suffering is the result of immediate sin. We have reason, indeed, for believing that there is in the case of man some real and mysterious connection between sin and sorrow; though our Lord here plainly implies that special suffering may be the effect of *general* sin instead of *special*.‡

But if we pass from man to the animal kingdom, we find clear proof of the existence of suffering and death in periods of the earth's history antecedent to the creation of man, antecedent, that is, to the existence of human sin.§ And though we are not absolutely warranted in extending to man the separation between sin and sorrow which we thus see existing in the case of animals, yet we should naturally infer from such fact an antecedent probability that human history would offer some examples where human suffering was not the effect of sin, but merely a continuation of that larger system of the permission of pain by Providence, of the operation of which, antecedently to human creation, we find positive proof. May

---

* Luke 13 : 4. † John 9 : 3.

‡ Compare Bishop Warburton's Sermon on the Lisbon Catastrophe (Works, v, pp. 286-298).

§ See the next Sermon, and the note at p. 89.

we not take the probable existence of such instances as an unexpected means of explanation and corroboration of our Lord's words? May we not adduce them, as in some humble degree explanatory also of the permission of woe in cases where we have no reason to infer the existence of a judgment for sin? Though we must speak with hesitation, and cannot hope to penetrate far into the purposes and plans of the inscrutable God, yet we may humbly and reverently venture to hope that we may gain by contemplating the Divine doings some trace of the possible cause of such permitted woes. Accordingly, in speculating upon great calamities like the Lisbon earthquake, we are compelled with reverence to answer, that the event was, to speak after the manner of men, an accident; that certain causes producing earthquakes are at work in the interior of our planet, and that those causes acted at that moment and in that particular spot.* More we cannot say. We cannot tell why some counter force was not benevolently operating to prevent it. We take it as a proof that the operation of general causes is not suspended by the Almighty, and occasionally not even checked by counter causes, but is still allowed to go forward, even when the continued effect of their action is the means of destroying sixty thousand persons who were not instrumental to produce the mischief, and who were powerless to avert it.

* If further proof of this position were needed, it might be found in the fact that where earthquakes have occurred in districts which are various in their geological character, the destructive effects of the earthquake have depended on the peculiarity of the strata on which the different towns lay. The places, *e. g.*, situated on crystalline limestone have been almost unhurt; those, on the contrary, which lay on clay or lava, have been been rocked into ruins. See Scacchi's account, "Del Monte Vulture e del Tremuoto ivi avvenuto nel anno 1851;" and Lyell's "Principles of Geology," ch. 28.

We might pass to other proofs that God mysteriously permits general laws to operate without interfering to check the misery which they inflict, drawn from accidents which are the effect of man's own imprudence and want of foresight, but which involve in their consequences those who are innocent of participation in the neglect which is their cause. Consider, for example, the terrific explosions which not unfrequently occur in collieries. Picture to yourselves one of those scenes. At the depth of hundreds of feet below the surface of the ground there exists a subterranean city, in which the coal rock is quarried by men living in a close temperature, supplied artificially with air and light, in presence of the constant development of a noxious and inflammable gas. Long galleries, diverging like the streets of a city, separate the miners by miles of tunnelled passages from the only outlet which exists in case of danger. In spite of all precautions on the part of the proprietors, some sudden act of imprudence occurs on the part of some miner. The gas ignites, overpowers the force of the artificial current of air, and sweeps, with devouring rush, through the close galleries of the mine. Many are instantly hurried into eternity; or, cut off in remote parts of the mine and unable to communicate with the surface, they die cruelly of starvation or are burnt by the fire, which, igniting the solid coal rock, turns those caves into a vast subterranean furnace.*

Is there no kind Being to aid those innocent men who die by accident or the imprudence of a fellow-workman? Is there no God of mercy to notice their unprepared souls about to be called to His judgment-seat,—none to see the

---

* This, it will be remembered, is a literal description of the accident at a colliery at Lund Hill, near Barnsley, in January, 1857.

life-long sorrows of children cast upon this wicked world fatherless; and of wives cast upon this cruel world widows? Yes! there is a Divine Being; but He is pleased to govern by general laws. And the general law that gas shall, under certain circumstances, ignite, is in His mysterious Providence allowed to have its course. The circumstance comes, the law holds on its course, and the catastrophe is its consequence; and the mode (as I shall presently show you) by which we reconcile such occurrences with the Divine benevolence is by supposing that the suspension of the general law would be a greater evil than that which ensues by its being allowed to have its course.

We might multiply illustrations, but it is only necessary to refer further to one event of deep sadness which will suggest itself to every mind. We may learn in the miseries that have befallen our eastern empire how true it is that the Divine Being in some respects leaves nations, as well as mere brute unconscious matter, to the operation of general laws. Let me not be misunderstood. I do not mean that the Indian revolt is purposeless on the part of Providence; but I wish you to separate between the moral lesson which men may derive from a calamity, and the final cause or purpose why the Divine Being has sent it. When the explosion has happened in the mine, we naturally gather a lesson of precaution against the recurrence of the accident, but we should not suppose that the Almighty had sent the explosion specially to lead us to improve the lamp and to rebuild the air-courses of our mine. Rather we should attribute the accident to a general law, and without pretending to fathom the motives or purposes of the Almighty, we should derive a valuable lesson from the occurrence. Similarly also in the existence of a panic in commerce. We ought to gather a lesson as to our own deportment in guarding against its recurrence, without supposing that Providence had sent us the woe simply to teach us this lesson.

The case is similar with respect to the miseries in India. It has been commonly asserted by irreflective minds that those woes have been sent as a direct visitation for the simple purpose of arousing slumbering England to evangelize Hindostan. This is right and logical if they mean that such may lawfully be our lesson from them in order to prevent their recurrence, just as the explosion of the coal-mine warns us to take measures against its repetition, or as the outbreak of a fever stimulates us to use sanitary measures.

I should be sorry if I were thought by my remarks to undervalue that blessed, that godlike work of missionary labor in which so many saints have won for themselves immortal honor. The names of Xavier and Schwartz, and Heber and Martyn, inscribed in the roll of the Christian heroes,—men of whom the world was not worthy,—who have perished in striving to evangelize Hindostan, would testify against him who should be impious enough to undervalue the work which they loved to the death. "I saw under the altar the souls of them that were slain for the word of God, and for the testimony which they held. And white robes were given unto every one of them. Therefore are they before the throne of God, and serve him day and night in his temple."*

Yet though I would not undervalue the moral lesson of increased missionary activity which is taught us by the Indian miseries, let us be careful not to confound this with the notion that no other object entered into the Almighty's purposes than the effecting this moral result, just as we

* Rev. 6: 8–11; 7: 15.

should avoid the confusion of supposing that His sole object in sending a fever is to lead men to attend to sanitary considerations. In each such case of an earthquake, or a fever, or a rebellion, we ought to distinguish the three following things one from another:—1st, the antecedent causes which have brought about the event; 2d, the moral purpose which the Deity may have had in sending or permitting it; and 3d, the moral lesson which man may rationally gather from it for his own conduct. Accordingly, when we pass from the moral lesson in each case which we may properly collect, and from speculating about the purposes of the Divine Being, concerning which we really know nothing, except when they are revealed to us by an inspired prophet, to examine into the causes, *i. e.* the antecedent circumstances from which each of such phenomena has arisen, we shall find them to be brought about by the uniform operation of fixed causes. The atrocities of the Hindoo rebellion are unfortunately no isolated phenomenon, but almost find their parallel in severity, if not in concentration (we regret to have to say it), in other centuries of the world's history. The mutiny of a pampered army under the combined influence of religious panic and frantic patriotism, at the suggestion of designing persons, is no isolated phenomenon; both alike have arisen heretofore from the play of human passion and human appetition, and will continue to arise unto the end of time; and mysterious as is the slaughter of hundreds of our innocent countrymen, we take that mystery to be but another proof of the wondrous administration of the Almighty by general laws. As an instance of an indiscriminate slaughter of guilty and guiltless, we place it parallel in the page of history with the devastation of the Palatinate in the 17th century; or with the massacre of the French which disgraced Sicily in the 13th; or that still more fearful atrocity, which stands out from among the many bloody deeds of the 16th as a monument of crime, the massacre of the Huguenots on the Feast of St. Bartholomew.

All, especially the last, are instances of a mighty slaughter permitted by a Providence which interfered not to stop those general laws which regulate human passion, nor to intercept those effects which the ingenuity of human sin produces. All alike are but the repetition, in political accidents, of the earthquake, or the explosion, or the pestilence in the physical. You may gather what lessons you please as to your future behavior in order to prevent their recurrence. But if you look to the cause of what is past, you find its explanation in that mighty wonder which we are wishing to impress upon you,—that causes which involve suffering are allowed by Providence to have their play, even though they involve the innocent in the sweep of their operation; that it seems true that in some regions of nature (if we may use the illustration without irreverence) Providence allows the world to move on, like some great machine,* which its author has set in motion as it were (to speak after the manner of men), but with some of whose wheels and movements he is not pleased afterwards to interfere. "Canst thou by searching find out God? Canst thou find out the Almighty to perfection?"

I have now offered a few illustrations of the absolute invariability of some portion of the Divine administration by law, even when such invariability is fraught with suffering to individuals. But I should be very sorry if I were to leave on your minds the impression that there is any degree of injustice, or any absence of benevolence in the permission

---

* Compare Babbage's "Bridgewater Treatise," ch. 8.

of these miseries, or that there is no real Providence in them. We are obliged to conceive of such events under the medium of human language and the illustrations drawn from human experience; and so I spoke just now of the world as one great "machine" which, as it were, acted by delegated power without the immediate operation of God. I meant not this to be understood literally, but only by way of explanation. When we speak of such uniform operations of general laws, we intend not to exclude the idea of God as working and omnipresent; we only express the uniformity of the system according to which He is pleased to work. Our finite minds cannot comprehend the operations of a Being whose government sustains the universe, any more than we can comprehend the attributes of His infinite mind. So, without doubt, if we could comprehend that infinite system, we should see that the catastrophe is not unnoticed by God, the material law not disconnected with the moral, natural accident and moral government not without their links of union.

And as I wish you not to carry away the notion that there is no Providence in catastrophes, so also you should not think of them as marked by injustice. In questions of this kind it is enough for us to rest in the fact that other and more comprehensive instances of Divine benevolence exist, which show that the *general* purpose of the scheme of nature is a benevolent one.* Our inability to compre-

---

* This principle of "the greatest happiness of the greatest number" has, it is well known, been adopted as the ground of morals in Bentham's philosophy. Pope, at an earlier period, not only applied it as the rule of human conduct, but as the measure of the Almighty's purposes, *e. g.*:—

"The universal Cause
Acts not by partial, but by general laws,

hend that scheme as a whole may well make us sure that if we could so understand it, we should see that these apparent exceptions are not such in reality. Just as if we stood looking on an ingenious machine, the general effect of which evinced consummate wisdom in its maker, we should at once think that any portion of it which seemed useless or injurious would have its use, if the inventor of it were to explain to us the plan of the instrument; so when we look on the great machine of the world or the universe, we may be sure that the apparent exceptions to a benevolent object in its construction would be seen to be reducible to agreement with the Divine mercy, if we could comprehend its scheme and its harmonies. Nay, the very idea of these apparent severities which I have attempted to convey to you, has been intended to remove any misgiving which might be felt in reference to them. For though we cannot hope to explain them fully, yet we have ventured to suggest a partial explanation, viz., that such apparent severities arise from the fact that the Almighty allows general laws to operate, and the very idea of a *general* law possibly excludes (as Bishop Butler observes) the idea of meeting all possible contingencies,* and implies that it must bear heavily in some special instances. We do not assert that

> And makes what happiness we justly call,
> Subsist not in the good of one but all." (Ep. 4 : 35.)

The writer of these Sermons does not wholly accede to either of these applications of it; but merely suggests that, in the absence of any better explanation of the mystery, we may lawfully adopt the principle in the kind of manner developed in the text, as a probable means of reconciling God's permission of suffering with the idea of benevolence in His character.

* Butler's "Analogy," part i, ch. 8, p. 132. See Brown's "Philosophical Works," vol. iv, Lect. 93, 94.

this is the case, but we put the supposition that, even if it be so, there is benevolence seen on the large scale even here; for the government by general laws is itself an act of benevolence.

We need only reflect for an instant on the amazing wisdom shown in some of these general laws and adaptations, in order to feel convinced that the wisdom is itself benevolence. A familiar illustration will explain my meaning, the use of which may be permitted, though I have made use of the same thought in this place before.* Go forth any fine evening and cast your eyes upwards to the stars scattered in glittering millions on the dark vault of the heavens. Though numerous as the sands upon the seashore, yet the movement of each star and each system is regulated by the most complete harmony. We know little of the vast number of those heavenly bodies; but of a few which lie nearer to our earth we know enough to be able to understand their movements and to predict their positions. And when by the engine of a refined calculation we compute their relations in distant time to come, we are allowed to understand the amazing wisdom with which the Divine Creator has guided their movements. For in estimating the mutual disturbances of the elements of their orbits, we are brought to conceive of a time in the distant future, when it seems that their perturbations shall exceed the conditions of stability, and cause an immense catastrophe. And is there really to be this catastrophe? No! at the very moment when we seem on its verge, we find that a series of compensations will commence, which will precisely

---

* The allusion here is again (as at p. 48) to Lagrange's problems. The mathematician will remark that rigorous precision of description is intentionally sacrificed for the purpose of clearness of illustration.

bring back the system to that state in which it existed before. The system has oscillated like a pendulum to that point, and then begins its backward circuit. The cycle of time required for that reversal of the oscillation (if we may so describe it) must be in some cases millions of years. So that we arrive at this stupendous result, that the Divine Being has impressed a simple law upon these heavenly bodies according to which they move; and yet this law is so exquisitely perfect, that He has by it anticipated the contingencies which will occur in the inconceivably distant depths of future time. "Lo, these are but a part of His ways. Canst thou find out the Almighty to perfection? It is as high as heaven, what canst thou do? deeper than hell, what canst thou know?" As you meditate on that consummate wisdom and prescience, is it not a proof that the government, by general laws, is itself an act of benevolence? The wisdom is the benevolence.

And as we see this truth in the physical, so also it holds good in the moral world. If I could not know the mind of God with regard to me, if He governed me by caprice, His conduct differing towards me to-day from what it was yesterday, I should not know how to deport myself before Him; but as He governs by general laws, His character never varying, from everlasting to everlasting unchanging, I accept that government itself as the best proof of His benevolence, because I see that it gives me a fixed principle by which to guide my conduct. If I sin, I know that there is no escape from the law which unalterably annexes punishment to the offence; if I obey God's laws, I know that their stability is the guarantee of my security. So that we can now not only say, "Great and marvellous are thy works, O Lord God Almighty;" but also, "Just are thy ways, thou King of saints; who shall not fear thee, O Lord, and glorify thy name?"

Here we must pause, omitting the consideration of the consoling doctrine of a Special Providence which is revealed in Holy Scripture. The necessity for this omission is to be regretted, because there is always harm in the exclusive attention to one side of a solemn question like this, and more especially when we are groping for truth without the guidance which the blessed Spirit of God offers us in His inspired word. I have wished to speak of this subject in a reverent manner, and though presenting strongly the secular view of Providence, yet I have endeavored to harmonize the proofs of God's power in universal law with His wisdom and benevolence. It is nevertheless possible that on such a subject our speculations may be wholly wrong; the views which I have given may involve some huge mistake. If such be the case, may God forgive them, and overrule the errors for His honor! We search for truth; but when we attempt to lift the veil of the spiritual without the aid of the voice from the unseen, we grope well-nigh in vain:—

> "As infants crying in the night,
> As infants crying for the light,
> And with no language but a cry."*

Yet I hope that we have arrived at a nobler and more cheerful view of Providence than those Stoics and Epicureans held, to whom St. Paul proclaimed that God is not far from every one of us. For we have wished to think of Nature's laws only as God's mode of working, and their invariability as the unchangeableness of His all-perfect

---

* Tennyson's "In Memoriam," p. 77.

government.* And I have failed to convey to you the meaning which I desired, if I do not send you home with the conviction that even the darkest dispensations of Providence are an equal proof of God's love with the brightest; that the sufferer in his deepest moment of gloom is as much the object of his Maker's care as when abounding in joy. The Divine Being sees that mourner, for He is not far from any of us; and if He sees wise in his Providence not to suspend the law which brings the suffering, that absence (to speak after the manner of men) of interference is not neglect. He is really bestowing his care. The sufferer is under the Providence of a personal God. Oh, it is a joyous thought that yon Englishwomen who were lately martyred for their country's honor in the far East were as much the objects of God's care, though He wrought no miracle to protect them from the fierceness of human passion, as their sisters who land on our shores daily with their tale of woe and their grateful hymn of deliverance. Yon soldiers, whose memory is so dear to us, who, in the assault of our enemy's stronghold,† paid with their lives for the noble prize of victory which their country has won, were as much the object of their Maker's care, though His hand warded not from them the stroke of death, as their comrades, whom we shall welcome back to our land, waving in triumph the colors which they proudly followed to victory. It is a joyous lesson to learn from this contemplation of God's general laws, that, suffer what we may, and die where we may, the suffering is not directed by chance. It is not inflicted on us capriciously; its infliction is a proof of love; for it is part of a great system which is guided by a Being

---

* See "Dialogues on Providence, by a Fellow of a College," a little work, very original and suggestive.

† Delhi, recently taken when this Sermon was preached.

all-powerful and all-loving. It comes from His hand. Though myriads of links in the chain of causation may separate us from Him, yet it is His act, His personal act, the expression of His all-perfect will, "for in Him we live, and move, and have our being."

Surely, brethren, under this light the consideration of general Providence has led us to the same result of a resigned spirit which the Scripture inculcates, and the same confidence which it inspires.

Nor is it necessary to add one word more, save to remind you that besides this general Providence of which we have spoken, there is another system taught us in the Bible (if indeed it be not rather in some incomprehensible manner a portion of the same),—a system perhaps in itself as general, yet suited to every need, directed by one who knows human wants; for it is administered by the God-man Jesus. Here we can take our refuge. When I think of those laws of absolute generality which Nature shows me, I tremble sometimes lest I may be overlooked; but when I remember that in Jesus there is a human nature mingled with the Divine, I feel sure that He is a being who knows what special wants mean, who can be touched with human sensibility, and can remember the woes and temptations of human infirmity.

What a blessed and amazing thought! Yonder on the throne there sits this God-man. Within the very shrine of the eternal glory, He has mounted up to plead for sinful men. Yonder, by the side of the Infinite One, who holds in the compass of His laws of infinite generality the infinity of the visible and invisible creation, is one conscious of our needs and touched with our infirmities.

Yes! we know that we are as much the object of that Saviour's mercy as though this universe were empty of all inhabitants but ourselves. He knows what we need. He cannot be perplexed by multiplicity, nor confounded by minuteness. Therefore we may leave all confidently in His hands, committing ourselves to Him in prayer; and though we may have to wait for the dawn of the eternal morning to illumine some of the dark passages of His Providence, yet we may rest confident of His power, His wisdom, and His goodness. He is omnipotent to save us, because He is God. He is willing to help us, inasmuch as He is man.

> "I cannot always trace the way
>   Wherein the Almighty One doth move;
> But I can always, always say,
>   That God is love."

Note

On Special Providence.

The following is a brief outline of the Sermon on this subject, which was designed to follow the preceding one:—

First, an investigation, conducted historically, into the teaching of Scripture on the subject, would have been given.

Then, a sketch of the schools of thought, in which the doctrine of Special Providence has been denied, with illustrations of their influence on literature, as, *e.g.* in the poetry of Pope.

Next, an investigation of theories, which have been supposed to suggest a reconciliation of the doctrine with the existence of general laws, such as (α) the Monadic theory of Leibnitz; with illustrations, showing how modern physical investigations, by resolving various supposed forms of matter into *power*, seem to lend support to something like his theory; and (β) the *machine* theory of Babbage's "Bridgewater Treatise."

After criticism on these attempts at explanation, it was proposed to examine whether the Scripture teaching must be surrendered, as merely a human or Jewish point of view; and to show that such is not the case, by offering tests to distinguish the human from the Divine element in the inspired teaching of Scripture.

Thus, assuming that we must believe with equal confidence in general laws on the evidence of Science, and Special Providence on the evidence of the teaching of Holy Scripture, it was proposed to examine this apparent paradox, investigating the ideas of Mr. Mansel (at that time only inferred from his metaphysical works, and from his tract on "Eternity," but now so ably exhibited in his "Bampton Lectures")—ideas which are in part an application of Kant's philosophy,—which would make such a paradox to arise from the incapacity of the human mind to comprehend such an object, not from real contradiction in the object known.

After this investigation, it was intended to suggest the possibility of a system of moral providence revealed in Scripture, as actual part of the system of physical providence, developed in Science, harmonious with it, and not contradictory to it.

Lastly, some notice would have been taken of the fallacy by which persons conceive of a general law, as if it had an existence apart from the individual instances which make it up. This fallacy, an offshoot of the ancient Realism, besets the human mind alike in its conception of general laws in nature, and of God's government in the Christian Church. The body of the Church has no more existence apart from its members than a general law has apart from the instances which exemplify it. Part of the confusion in regarding laws of Nature as being distinct from God's working in Nature, seems attributable to this fault of giving real existence to human generalizations.

The Sermon was not preached, partly because of the long interval of many months which intervened before the opportunity occurred for it; and partly because, in the meantime, Mr. Mansel, in the 6th of his "Bampton Lectures," had sufficiently investigated the subject and preoccupied the ground.

SERMON III.

DIVINE BENEVOLENCE IN THE ECONOMY OF PAIN.

(PREACHED BEFORE THE UNIVERSITY, FEBRUARY 13, 1859.)

GENESIS 47 : 8, 9.

"*And Pharaoh said unto Jacob, How old art thou? And Jacob said unto Pharaoh, The days of the years of my pilgrimage are an hundred and thirty years; few and evil have the days of the years of my life been, and have not attained unto the days of the years of the life of my fathers in the days of their pilgrimage.*"

THESE words record a scene which thought might well love to dwell upon, and art to depict, even if the lesson to be learned from the view of life contained in them were less valuable than it is.

The scene is a striking one,—the interview of a Hebrew shepherd, chieftain of the desert, with the haughty Pharaoh, monarch of the first empire of his time. It carries us back to an age of the world which it is hard to realize in thought, which has almost left no traces of its power in the remains of its public works, and seems well-nigh to live alone in the interesting narrative of the Book of Genesis. We are accustomed to reproduce to ourselves the image of the greatness of Egypt by reconstructing in our minds, and repeopling with their ancient proprietors, those temple palaces whose gigantic porches, or curious columns, or beauteously-graven obelisks still adorn the banks of the Egyptian stream. Yet all these, old though they be, are subsequent to the age of Jacob. In order to conceive of the Egypt which Jacob visited, and of the Pharaoh to whom Jacob was introduced, we must go back in thought to a time still older, a period when the art of carving obelisks and of erecting porticos was yet unknown,* and when the people passed their lives in houses of wood, and entombed their ancestors in massive pyramids, which outlive the changes of nearly forty centuries. Those pyramids, which now look down in gloomy magnificence on the desert scorched into barrenness around them, around whose massive bases hardly a sound of animate life is now heard to break the everlasting silence of the desert waste, stood (it is now understood from their inscriptions, in spite of the opinion of the Greek historian†) in the days of Jacob as they stand now; and if we measure their magnificence, and then strew the desert plain with the traces of active industry and the bustle of a thriving population, we may be able to form to ourselves some notion of the scene which must have presented itself to that patriarch when he came into the country over which his son Joseph was minister; we can

---

* The obelisks and porticos chiefly belong to the great age of Egyptian art of the 18th and 19th dynasties. See Fergusson's "Handbook of Architecture," vol. i, b. v, ch. 1, 2.

† Herodotus (ii, 125) attributes the erection of the Pyramids to a period hardly earlier than B.C. 1000. The monumental evidence shows that they were the work of the 4th dynasty, which, according to the most moderate computation, must be some centuries before the time of Abraham. See Rawlinson's "Herodotus," vol. ii, ch. 8, p. 344, &c.

reconstruct from these fragments some idea of the Egypt of Jacob's day.

As the relics of Egyptian architecture enable us to picture to ourselves the Pharaoh who was one of the characters of the interview narrated in our text, so the unchanging characteristics of the shepherd life of the Arabian and Mesopotamian deserts reproduce to us the external features of life and manners of the Hebrew patriarch who was ushered into the Pharaoh's presence. We can imagine to ourselves the bearing of the shepherd chieftain, accustomed from childhood to the wandering pastoral life; his head hoary with age; his countenance bronzed with exposure to weather, and furrowed in deep lines, which told with unmistakable clearness their tale of hardship and trial; his manner dignified by the conscious self-respect which belonged to one who had long been the chieftain of a potent tribe, carrying the modest but manly consciousness of the liberty of a child of the desert, even in the servile court of the Egyptian autocrat.

Such is the scene. The city, perhaps, of Memphis; the court of a Pharaoh, surrounded by his attendants on the one hand, and the venerable shepherd patriarch on the other. A son of that old shepherd, now prime minister of the Egyptian kingdom, himself long ago transformed, to all appearance, into an Egyptian, in every respect, save in the filial affection for his ancestry which still throbbed within him, disdains not to introduce that old man into the sovereign's presence. Let us listen to the interview: "And Joseph brought in Jacob his father, and set him before Pharaoh, and Jacob blessed Pharaoh. And Pharaoh said unto Jacob, How old art thou? And Jacob said unto Pharaoh, The days of the years of my pilgrimage are an hundred and thirty years; few and evil have the days of the years of my life been, and have not attained unto the days of the years of the life of my fathers in the days of their pilgrimage. And Jacob blessed Pharaoh, and went out from before Pharaoh."

There is something very natural, very fresh, in the words which Jacob used, "The days of the years of my pilgrimage." They were precisely the idea of life which would present itself to one accustomed to no regular home, but wont to move his encampment from spot to spot to find pasturage for his flocks and herds. Life would seem to him eminently "a pilgrimage," a sojourn. Also, the complaint, "Few and evil have the days of the years of my life been," is just the kind of plaintive, melancholy utterance of an aged man, with life behind him, a scene of sorrow, and nothing but death and gloom before him. They quite express the kind of regret which an old man would feel, the retrospect which in all ages a thoughtful mind would take of its past life; but which would come forth especially from a man like Jacob, whose life had been unusually chequered with evil,—evil which he had done, evil which he had witnessed, evil which he had suffered. And we can well imagine that in those long years of solitary sorrow, in which he had mourned the entombment of his earthly happiness, when he had buried his loved Rachel beneath the pillar in Ramah, and the sad end, as he supposed, of his son Joseph, one of the two children which we learn from the narrative seemed to him the relics bequeathed to him from their lamented mother,* he had employed the leisure of a shepherd's life and the inactivity of age in the sad but serious view of life which found its instinctive utterance in Pharaoh's presence; when, in answer to the monarch's

---

* Gen. 35: 18; 42: 36; 43: 14.

question, "How old art thou?" he was unable to reply without leaving on record, in words whose plaintiveness touches us even at this distance of time, his sad experience of human life, "Few and evil have the days of the years of my life been."

We have now dwelt, I should hope, long enough in thought on that ancient interview to realize it vividly to ourselves, and to enter into the feeling expressed in the utterance of the aged patriarch. But what religious and moral lesson may we learn from it? The one which I wish to draw is this, to fix the mind on that idea of life which Jacob here expressed. In his retrospect of it, evil and sorrow seemed to outweigh joy and pleasure; the balance was in favor of gloom.

How completely is the experience of life which the patriarch draws from his own personal history confirmed to us, by the testimony which the subsequent history of the earth has unfolded to us; the varied events of woe that have arisen in the development of the world's mighty drama since that early time! What is the voice of history but a roll written within and without, with mourning, lamentation, and woe? What is it but an illustration of the great fact, that Providence allows human life to be marked by agonizing sorrow? How many evils exist, brought about by men,—wars, revolutions, cruelty? How many permitted by Heaven,—poverty, famine, disease, bodily infirmities, the catastrophes of accidents, sudden deaths? What can more truly describe the feeling of the mind, which looks upon the world's history from this point of view, than to exclaim of it as Jacob did of his own life, "Evil have the days of the years of its life been?"

Now, what is the cause of this permitted pain? and how can the existence of such manifold misery be reconciled with the idea of a benevolent character in the Divine Creator? It is a very small portion of this great subject which can come under our notice on the present occasion; yet I hope that a few considerations on the use of Pain will at once excite in us a reverent feeling towards the Divine Being, whose government we shall perceive to be guided by mercy, even in its forms of terror, and will stir us up to a true perception of our own duty alike to our neighbors and to Him.

The ideas which I wish to bring before you are these,— that though we cannot entirely fathom the mystery which is involved in God's permission of pain or suffering, yet we can discover in it proofs of His mercy, not merely in the very purpose* of its administration, but in the twofold remedy which He has provided for its diminution, in the progress of civilization and in the mission of philanthropy and of Christianity.

It will probably occur to many of you, that pain is the effect solely of sin, and therefore that whatever woes mankind may suffer under the economy of it, are brought on by their own fault. It is indeed true that much of pain is the effect of sin, and would never have existed if sin had been absent from the earth; but this does not solve the whole mystery; for there is much pain, it is now clear, which is not the effect of sin. The economy of suffering is a far grander thing, is part of a far grander scheme of God's administration, than we are at first led to suppose. Providence has shown us, by the discoveries of Science in

---

* Similar proofs of benevolent arrangements can be shown in the very *nature* of its distribution. But the inquiry was too physiological to be introduced into the Sermon. Mr. G. A. Rowell, of Oxford, has treated this particular aspect of the subject in an interesting essay on "The Beneficent Distribution of the Sense of Pain."

the present century, the mysterious fact, which we should not otherwise have suspected or guessed at, that pain and death existed before the creation of man,—before the existence of human sin.*

It used to be conceived that, about six thousand years ago, the Almighty's creative fiat first broke in upon the stillness which existed in universal nature, and evoked from nothingness this globe, and strewed the sky with the orbs which are scattered in glittering millions, and decked this earth with plants, and peopled it with animals for the use of man. It is now known that this opinion is not correct, and that the narrative which was supposed to tell us so, can at most refer only to the preparation of the earth for the use of man, and not to its original construction. The first origin of creation must be placed back at a period indefinitely remote. Through a succession of ages and cycles, the profusion of God's creative hand gave life to myriads of species of animals and plants before His boundless love suggested the thought, "Let us make man after our image." The science which has explored the rocks of the world has deciphered in them the written history of God's government of this planet in ages upon ages anterior to human history. The whole earth is one huge sepulchre of the remains of former worlds. The marvellous fact upon which I am wishing now to dwell is this, that in those ages when man was not, and when the fish of the sea, or huge reptiles of the marshes, or the gigantic mammoths of the

---

* It will be observed that the truth of the teaching of geological theory, in reference to the occurrence of death antecedently to the creation of man, is here assumed. Indeed it can now not be doubted by any educated person. A note is, however, appended to the present Sermon, to dissipate some objections which are still taken against geological science.

forest, were the sole lords of the planet which now forms man's habitation; when accordingly there was no sin, because the irrational animals were incapable of sinning, yet pain existed there and death likewise,* and those great physical catastrophes, such as earthquake and volcanic eruptions, which destroy animal life, were also abundant.†

---

* Though we may rest unhesitatingly in this truth, proved by irrefragable evidence, and may feel sure that the method of reconciling it with previously known truths will hereafter suggest itself, yet, as many conscientious men feel a difficulty in accepting it in consequence of its contradiction both to St. Paul's statement (in Rom. 5 : 12), "By one man sin entered into the world, and death by sin; and so death passed upon all men, for that all have sinned;" and to the statement in the Book of Genesis (3 : 17), "Cursed is the ground for thy sake," it seems fair to them to enumerate some of the modes which have been suggested for the reconciliation of the discrepancy. These modes are by supposing: (1st), that though death belonged to the *animal* kingdom before the existence of sin, yet its extension to *mankind* was a judgment for human sin; (2d), that the pre-existence of death and deterioration was arranged by Providence, with a view to the future existence of human sin, foreseen by the Divine prescience; so that the world, according to this view, was really prearranged for the residence of *fallen,* not of *pure* beings,—an idea to which St. Paul's hint, that "the Lamb was slain before the foundation of the world," might be supposed to lend countenance; (3d), that as St. Paul is employing a process of argumentation in the passage cited, the same weight of inspiration need not reasonably be assigned to his arguments as to his positive authoritative statements; inspiration, according to this view, residing in the elevation of the *intuitional,* not of the *logical* faculty; (4th), that St. Paul may be regarded as merely repeating and reasoning on the Jewish view, according to the best information possessed at that time, before God had taught to man, through the revelation of uninspired science, a grander truth on this subject than He had vouchsafed to communicate to the Jews through the revelation of inspired messengers. This latter view would not deny the authority of Scripture, but only imply degrees of inspiration in its teaching.

† The destruction of animal life by earthquakes, such as those which

To adduce only one instance as proof. In many spots of the earth slabs of rock are dug up which contain the remains of delicate fishes which existed long antecedently to human history; their beautiful little forms being stamped upon them, still contorted in the agonies in which they expired.*

What do these facts teach us? They reveal to us this amazing truth, that the economy of pain, which we had thought to appertain to man, and to be the effect of sin, is part of a much larger scheme of Divine Providence, extending backwards to times of which we had no conception, and designed for purposes larger than we had imagined. Science has in fact, in this case, become a revelation.† It has advanced our knowledge, not only of Nature, but of the system and purposes of God in ruling Nature; and I have ventured thus to allude to it, because it is most desirable that the minds of our students should be freshened by acquaintance with the discoveries of Science, and that they

---

raised mountains, or produced the dislocations usually called "faults," is an inference; but that which was produced by volcanic eruptions is a fact proved by the existence of molluscous remains in the tufa of volcanoes of the tertiary age, both in central Italy and in Auvergne. The existence also of such catastrophes as sudden outbursts of poisonous vapor in the ancient seas, is the most probable supposition for explaining the aggregation of fossil remains of the same family, as *e. g.* of Belemnites, in some parts of the Lias; as if a shoal of fish had been destroyed by some sudden cause in the ocean, and entombed in its depths.

* Fossil fishes are frequently found; but it is in the remains of those discovered at Solenhofen, in Bavaria, that the contortions are most distinctly marked. They are found there in a schistose limestone, probably cotemporary with the upper oolites of our own country. See Lyell's "Manual of Geology," ch. 20.

† Compare the remarks on this subject in Sermon I of this volume (pp. 43–48).

should not go forth into life to propagate errors which are exploded among the educated, nor should receive the first information of their mistake from the harsh satire of some stubborn critic. Let us rather hail Science as a handmaid to religion. The inspired Bible is the revelation of God's scheme of mercy in Christ; uninspired Science is a revelation of God's majesty in Nature, surpassing in this respect the former, in unfolding the mightiness of His ways, and in enlarging our conceptions of the infinity of His purposes. And, therefore, we may take the facts to which I have alluded, as a proof that though some pain is doubtless the effect of man's sin, yet the government of God by pain is part of a wider scheme, of which, perhaps, we can hardly suspect the purpose.

So far as we are able to guess at its object, we may assert that the economy of pain is an economy of discipline. It appertains to a being that is in a state of progress; and so, instead of seeming to be cruelty, it is really mercy, because it is a lesson inculcating prudence and inducing improvement. Two instances will illustrate this.* If the little shell-fish which enjoys its life in the warm waters of a tropical clime possessed no sense of biting pain at the presence of cold, what should hinder it from allowing itself to be drifted by the ocean's currents to those colder waters for which its organism is unsuited? If the wild beast of the forest felt no sense of pain as its hairy skin is lacerated by the sharp branches of trees among which it rushes, what should prevent it from consummating the destruction of the very covering which was intended to protect it against alternations of climate? The endowment of pain is a real

---

* These two instances are borrowed from Mr. Theodore Parker's Sermons on the "Economy of Pain." (Sermons IX and X.)

kindness; it is the sentinel to warn against danger. If it be a punishment, it is only intended as a lesson against future imprudence, against the recurrence of the conduct which produces the pain. We claim it, therefore, as a proof of God's mercy that he has thus imparted to sentient beings a beacon to warn them against peril; and we cannot but suppose that it must have been some purpose of this kind, which was intended by the distribution of pain in those early ages of creation to which allusion has been made.

We might extend to the case of man the illustrations drawn from the lower orders of the animal kingdom. When, however, we thus pass from the merely sentient animals to the consideration of beings possessed of a higher nervous organization, and endowed with the attributes of reason, conscience, and responsibility, we naturally, as we should expect, find the capacity of feeling pain to be vastly enhanced, but designed with the same purpose of mercy. For the pain is commensurate with the discipline; it is a signal warning against harm or wrong; and as the discipline of man is more extended, his powers greater, his means of wrongly acting enlarged, so his capacity of feeling pain is also extended. He not only feels it in body, but experiences the pangs of mental misery, the lashings of remorse, the tortures of conscience. Yet these are mercy. They are all designed to deter from the repetition of the imprudence or the sin; they are intended as a warning to others who see the effects, that they may learn a lesson by example, without having to buy instruction through their own experience.

When the imprudence of a mariner dashes the vessel on the rocks, or when, through the neglect of common precautions as to health, a pestilence fastens on the plague-spot of a city, or the explosion of some dangerous factory maims or massacres scores of unoffending bystanders, there seems at first no mercy in these judgments of pain inflicted on the innocent; but when we consider what lessons they are intended to teach, they too are in their tendencies really mercy. It is only severe lessons like these which arouse men from a motive of personal safety to remedy the evils which imprudence or neglect has created. Thus, even in chastisement there is mercy; even in the dark and mysterious economy of pain there is proof that a God of love is ruling.

Yet the remarks which have been made apply only to that pain and suffering which is remedial; what shall be said of the amount of suffering permitted in God's providence, which comes upon man by no fault of his own, and which he is powerless to avert? A city lies sparkling in beauty: suddenly a low rumble is heard; it grows louder as it approaches; and when it is at hand, the city rocks like a ship laboring in a storm; the buildings crumble into heaps; and the glorious city, which a few minutes before was busy and bright with life, is a mass of ruins, with thousands of its population buried in its fall. The patient mariners, after braving many a danger, are in sight of the haven where they would be; but a storm of the ocean overtakes them; the vessel founders; and their last breath of agony is heard gurgling on the surface of the deep as they sink into its abyss. The toiling collier shall be digging in his subterranean city: some accident ignites the inflammable gas which issues forth from the coal-rock; the flame sweeps with devouring rush through the close galleries of the mine, and strews those dark caves with the corpses of innocent sufferers. The Divine Being is pleased not to suspend His general laws; and the general law that, under certain circumstances, the earthquake, or the shipwreck, or the explosion shall occur, is mysteriously allowed to have its course. The contingency comes, the law holds on its course, and the catastrophe is the consequence.* What shall we say of these permitted horrors? Can we reconcile them with the idea of the government of a God of love? We can in some sense do so,—at least so far as we finite beings can hope to comprehend the thoughts of the Infinite Mind. We assert that benevolence is seen on the large scale even here; we claim that the government by a uniform system of general laws is itself an act of benevolence.

It is not necessary to enlarge on this subject, because, on a former occasion, I endeavored to harmonize the apparent severity of such a plan of government with the idea of benevolence; and, with the view of showing that the amazing wisdom exhibited in its construction is itself kindness, I drew an illustration from some of the discoveries of mathematical astronomy.† I attempted to make it plain, that through calculations conducted by the instrument of a refined analysis we arrive at this marvellous result, that the Divine Being has impressed a simple law of such exquisite perfection upon the heavenly bodies, that He has by means of it anticipated the contingencies which will occur in their disturbances in the inconceivably distant depths of future time. Are not such consummate wisdom and prescience a manifest proof that the government by general laws is an act of benevolence? The amazing wisdom is itself the benevolence. Thus, though we cannot understand the whole mystery of these catastrophes which arise in the operation of such a plan of administration, we may be sure

---

* The ideas of the last few lines are repeated from Sermon II.
† See Sermon II, p. 75.

that general happiness is produced by the arrangement, in spite of occasional pain. The earthquake, or the shipwreck, or the explosion, produces misery, but the general system of wind and weather, and gas and air, diffuses general enjoyment; and therefore, even apart from their moral value as lessons, even on the physical ground merely, we can show that undeserved pain is compatible with the administration of a God of love. The means are severe, but the end is beneficent. "Therefore, hearken unto me, ye men of understanding; far be it from God that he should do wickedness; and from the Almighty that he should commit iniquity. Yea, surely God will not do wickedly, neither will the Almighty pervert judgment."

We have thus learned in the survey alike of the pain which is a warning to deter from harm, and of that which is permitted to occur in the ordinary operations of Nature's laws, that even the dark dispensations are an equal proof of God's love with the brightest, that even the mysterious economy of pain is an evidence of God's benevolence.*

But we have not yet exhausted the proof of God's mercy in this dispensation of severity. We have, indeed, seen its benevolent purpose and tendency; but we should also take into account that God has been pleased to institute two agencies which especially tend to diminish pain, and make it effect the moral purpose designed in it. One of these agencies is the benevolence which is called forth by civiliza-

---

* Pain seems to be various in origin. (1), It arises from the operation of general laws; (2), it is corrective; (3), it is designed to test character, as in the case of Job; (4), it is perhaps occasionally retributive. The two former branches only have been discussed above. Some hints for the discussion of the third may be found in an able article on the Book of Job in the Westminster Review for October, 1853.

tion; the other, the philanthropy which takes its rise in Christianity.

It may create a momentary surprise to hear of the relief of pain being the effect of civilization; for experience so often compels us rather to associate the idea of selfishness with that acquisition of wealth which marks a growing civilization, and heartsickening despotism with political centralization. Yet we shall perceive that it is so, if we look at two features which appertain to civilization, viz., the development of medical science, and the growth of public opinion. It would be so easy to prove from history that the increase of civilization has favored, nay, necessitated the growth of medical science, that we may assume the fact; for my object now is rather to regard the art of healing as an instrument in the hands of Providence for the alleviation of pain. We are accustomed frequently to take only a utilitarian view of it; when, however, we regard it from our present point of view, we must look upon the humblest practitioner, in his humblest employment, as an unconscious instrument in erecting one stone in the great temple, which God is building, of human happiness and human improvement. And we must also look upon the remedies which have been of such inestimable blessings to mankind, such as the febrifuge power of Peruvian bark, the practice of vaccination, the use of anæsthetics, though in themselves, as it were, accidental discoveries, yet in a higher sense as gifts of God, as merciful arrangements of His Providence for the mitigation of suffering. Though they be fortuitous discoveries, yet if it be the special prerogative of human civilization to acquire such knowledge, and if it be the inseparable quality which God has given to man to attain to civilization, we are not wrong in claiming them as evidences of the government of a God of love.

We may see a similar proof also, if we look at the trait of a mature civilization which is seen in the growth of public opinion, and its necessary concomitant, a free press. If we turn our thoughts to the state of our own country at this moment, we perceive that there is not a wrong, nor a supposed wrong of the most insignificant kind, which fails to excite through the free press of England, public attention and sympathy. The event may be in itself slight, yet it is felt not to be a trifle, because it involves a principle. It may be some bodily hurt of a poor person, or it may be some insidious attempt to obstruct public progress, or to sap the foundation of our liberty and our national independence; in either case Englishmen take it up, because they feel that the trouble is theirs. If the one member suffer, they know that all the members suffer. The hurt at one extreme limb of the body politic is telegraphed throughout the whole of its mysterious organism, and each sinew and each nerve beats responsive to the pain impressed on the distant member. The very exaggeration, the occasional abuse of this public sympathy, proves its power and its value. If here, again, we see that public sympathy is the effect of freedom, and freedom the effect of civilization, and civilization the gift of God's general providence, let us not omit to recognize in every act of sympathy which responds to the complaints of suffering, the pulsations of the personal will, the expression of the mind of love, which directs the first links of that chain, some of whose windings, as we see them in the tangled mass of human society, we have been attempting to trace.

Yet it is not merely in the benevolence of a growing civilization that we notice the merciful arrangements of Heaven for the mitigation of pain; we trace it much more in the mission of Christianity. Our holy religion is the reflection of the mission of its Divine author; He has left us an example that we should follow in His steps: and His mission was one of universal mercy, not to soul only but to body. In His journeyings over Judea, wherever He saw misery, physical as well as moral, He scattered it by the breath of His miraculous power. He bore our sickness, and carried our sorrows. And as He acted so did also His Apostles. From the very moment when their souls were baptized with the Pentecostal gifts, so that they understood what Christ's atoning death had wrought for them, and felt the holy love of God and of man stirred up within them by His Holy Spirit, they hastened to go forth on their mission of love. And not Apostles only, but humble members of the Christian Church counted it their highest privilege to minister to their fellows. So also, as the circles of Christian influence widened, institutions unthought of by heathens, were established for the relief of sorrow and the mitigation of suffering. The statesmen and monarchs of the ancient world constructed many works of public utility, but none directly adapted to the cure of disease. The ruins of their aqueducts still span wide valleys with their gigantic arches; their baths for the poor,* now crumbling in ruined majesty, form some of the most collossal and beautiful remains of the Eternal City; but no philosopher was ever led by his science, no statesman by his generosity, to construct hospitals, or a system of relief for the diseased. It was when He who had suffered as a man sent His Spirit down to melt the hard hearts of men into overflowing love, that the thoughts of visiting and assisting the sick first entered into the hearts of men to conceive.

And in various ages the most conspicuous examples of

* *E. g.* those of Caracalla and Diocletian.

heroic and enduring self-sacrifice have appeared in the muster-roll of those who have endeavored to carry out the secular mission of Christianity—its relief of pain. We need not go back to past times, and recall the memory of a Borromeo ministering to the population of Milan when smitten by pestilence, nor a Vincent de Paul sending forth the missionary sisters into ravaged Lorraine, nor a Howard, in his circumnavigation of charity, collating the distresses of all men. Our own age, our own memory, will supply to us conspicuous instances where practical Christianity has fulfilled its mission of plunging into the infection of hospitals, and diving into the abodes of sorrow. There is one recent scene in our national history which finds its place in the annals of the Christian mission of mercy; there is one spot on earth whither the philanthropist may take a pilgrimage to kindle his own energies. And as he gazes on the hillocks which mark the last resting-place of English heroes, and drops his tear of sympathy to the memory of those who bought with their lives the noble prize of victory for the country which they loved so well, his sympathy must kindle into intense energy as he turns to gaze on that huge square edifice that overlooks the silent cemetery, which has been consecrated by the presence of those Christian heroines who bent over the pillow and soothed the last moments of their countrymen who expired in the plague-struck hospital of Scutari.

Here we may see how truly Christianity still carries out its mission of healing; and in the spirit of this example we may fitly conclude our subject: for we have traced in the economy of pain the proof of God's benevolence, not only in the suffering which is the punishment of imprudence, and in that which comes upon us without our own fault, but also in the perpetual system which He has provided for the mitigation of sorrow in the benevolence of civilization and the philanthropic mission of Christianity.

Nor can this view have failed to exhibit to us our own duty alike to our neighbors and to ourselves. We must have felt that in mitigating the slightest pain in the most insignificant creature of God's sentient creation, we are co-workers with God, we are doing our part in the system in which He has placed us; in diminishing human suffering in the least degree, or adding to the stock of human happiness, we are following the footsteps of Him who Himself was the great example of the dignity of condescension, of the majesty of sympathy, of the divinity of pity.

It is a very cheering circumstance that the Christianity of our age is becoming awake to this its secular mission,— its mission to the bodies of men as well as to their souls, its mission of civilization as well as its errand of religion. It was the lesson of this kind which all felt that they had to learn, which not long since bespoke the sympathies of admiring England for the labors of the missionary explorer,* who, after receiving from his grateful countrymen his well-merited honors, has gone back with the unaffected simplicity which was the sweetest trait in his noble character, to bestow his labor of love in carrying up the streams of the African continent the seeds of incipient civilization, as the pioneer of industry, the harbinger of the bright day of improvement which shall in distant time spread its refreshment over the arid plains of the African continent. When that first laborer shall have passed to his reward (distant, God grant, may be the day), his name shall be blessed; though he may rest from his labors, his works shall follow him.

---

* The Rev. D. Livingstone.

But his example ought to animate us at home. For if we would find barbarians outside of the pale of civilization, and beings degraded below the level of humanity, we have no need to go to search for them among the roving Bosjemen of the Calahari desert,* we may find them nearer home, in the crowded English cities, amid the lazzaroni of our metropolis, amid the hopeless, homeless outcasts of yon great London. And if we would reach these with mercy, we must first feed them; if we would Christianize them, we must first unbrutalize them; if we would reach their souls, we must begin with their bodies; if we would hope to teach them religion, we must accompany it or precede it by attention to public health and comfort. Tracts, and Bibles, and clergy will do little, unless we afford also fresh air, and clean water, and wholesome food, and warm fuel, and healthful recreation, and the commonest rudiments of God's blessed gift of civilization. We must sacrifice for once those political practices (true though they are in the main) which the teaching of Malthus caused to be embodied in public law; we must come forward as a nation to help those who cannot and will not help themselves. We must purify the public drainage, and erect public lavatories, and build decent cottages wherein the sanctities of domestic life may find a shrine, without the herding together of persons, like brute beasts, without respect of age, or person, or sex, if we would wish to use God's method of elevating men, or would desire our religious efforts for their good to be anything but a mighty, heart-sickening mockery.†

---

* Livingstone's "Travels," ch. 5.

† It is undoubtedly true that under ordinary circumstances Christianity precedes rather than succeeds civilization. Religion begins from within and works outwards, first making the heart right, and then afterwards the life. Missions civilize by the very act of Christianizing.

Yet while we are learning this lesson of duty to our neighbor, let us not omit to learn also one in reference to ourselves. While we are laboring in our sphere to diminish human sorrow as well as human sin, let us not fail to realize the deep lesson which the sight of that sorrow ought to teach us, of unworldliness and of preparation for the future world. Let us not fail to feel that this life is verily a pilgrimage, a sojourn; that we are placed here to seize the few moments to prepare ourselves for another world, and that the evil of this life may be our very best preparation for the future, if only we are victorious through the help of Him who has loved us. We must learn to feel that here all is fleeting, there all eternal; here the shadow, there the substance; here the dream, there the awaking; here all marred and imperfect, and unable to satisfy the deep cravings of our immortal souls, there all blessed perfection, and God and goodness as the everlasting fountain and satisfaction of our intensest appetitions. Then we shall use the world without abusing it, and live here as heirs of immortality. And the strength of that conviction will make us tremble, lest, when we have entered on that other state of being, when return to this life is impossible, when our souls are stamped with an everlasting destiny, when he that is unholy must be unholy still, we should find that we have made an everlasting mistake, that we have allowed ourselves to be cheated by the dream of life, hurried away by its gaieties, bound down by its business, and have neglected to use its opportunities to secure a fitness for a home above. Let us learn this lesson, and carry it out in our lives. While we consecrate our efforts to bless our fellows, let us

---

But in extreme cases of degradation, such as abound in our larger towns, religious influences are rendered abortive unless assisted by civilization.

gather ourselves in the secrecy of earnest prayer to our common Father, and seek that He would keep before our souls the vision of the eternal world, and make our lives the means of preparing us for it. Let us ask the mercy which is free as the air we breathe to all who ask it in the merits of Christ, let us crave His help against sin, and His favor, which no one ever yet asked in vain. Then, if we do so, we may well hope that, as the evening of life closes around us, and we are ready to lament with the ancient patriarch, "Few and evil have been the days of the years of my life," we shall, in the recollection of a life well spent, catch a prospect in the future, a bright home beyond the dark valley of the shadow of death; and that with consolation cheering us such as fell upon him in his last moments, our souls may pass away from earth with the joyous thought, "I have waited for thy salvation, O Lord."

> "Life, I repeat, is energy of love,
> Divine or human; exercised in pain,
> In strife, or tribulation; and ordained,
> If so approved and sanctified, to pass
> Through shades, and silent rest, to endless joy."*

NOTE

On the Evidence of Geology.

It is hardly necessary, considering the manner in which a knowledge of Geological discovery now enters into the education of all cultivated persons, to add remarks on the irrefragable character of the evidence of those discoveries; yet some objections to them deserve notice, which

---

* Wordsworth's "Excursion," b. v, end.

exist in the minds of those whose reverence for old truths inclines them to adopt any excuse for declining to accept new ones.

These objections are (1st), that the phenomena of fossil remains can be entirely explained by a general deluge, without assuming the existence of death antecedently to the creation of man; and (2d), that Geology is so young a science, and has so often changed its theories, that hesitation in accepting its present teaching is excusable.

The former position cannot be held any longer by any one who will put himself to the trouble of examining conscientiously the steps of Geological proof; indeed, the persons who in future assert it must abdicate their claim both to impartiality and intelligence.

The latter position, though more plausible, is equally fallacious. The cause why Geology has changed its theories is, that the discoverers of the science were so conscientious, so afraid to draw inferences hastily which would clash with received beliefs, so unwilling to admit the new truths which God was teaching them through the revelation of Science, that they adopted premature attempts to adjust old beliefs to new discoveries. Accordingly, from time to time they were compelled to throw away some element in their conclusions, which fresh investigations showed to be no longer tenable. The changes in the theories of Geologists have not been those of men who were guessing at random; they have been the uniform progress of minds who had humility enough to lay aside their preconceived hypotheses before the newly-opening visions of truth.

In reference to the allegation that Geology is a young science, it should be remembered that since the establishment of ascertained methods of investigation and of proof, a science constructed upon such methods possesses immediately the certainty of older sciences, the larger portion of the history of which has only been the random attempts at discovery, which were made antecedently to the establishment of correct methods.[*] Bacon said that the method of science would grow together with the sciences,[†]—a remark which experience has confirmed. Men have, as it were, stumbled upon discoveries, and having done so, they have turned back and read in those discoveries the theory of the method by which they attained them. They have read in science the logical

---

[*] This is exhibited clearly in Dr. Whewell's "Philosophy of the Inductive Sciences," in the chapters where he traces the gradual evolution of scientific ideas; and in Professor Baden Powell's "History of Natural Philosophy."

[†] Nov. Org. B. I., *in fin.*

method of scientific discovery, and hence the modern inductive logic of scientific method, as shown in the great modern writers on the subject,* is itself a strictly inductive science, a rigorous statement of the methods which have led to the verified discoveries in the sciences.

Hence the allegation that Geology is an uncertain science, because a new one, disappears, inasmuch as it is a science founded on ascertained methods; indeed, such a charge is as absurd as if a person were to object against some modern astronomical calculation that it has been executed too quickly, because the astronomers who lived before the perfection of analytical methods of investigation would have taken much longer time in the discovery of it.

* Sir J. Herschel's "Introd. to Nat. Phil.," Part II; Dr. Whewell's "Philos. of Induction;" Ampere's "Essai sur la Philosophie des Sciences;" Comte's "Positive Philosophy;" Mr. J. S. Mill's "System of Logic."

SERMON IV.

JEWISH INTERPRETATION OF PROPHECY.

(PREACHED BEFORE THE UNIVERSITY, FEBRUARY 24, 1856.*)

ISAIAH 6:9.

"*And he said, Go, and tell this people, Hear ye indeed, but understand not; and see ye indeed, but perceive not.*"

THESE words were spoken in the marvellous vision which was vouchsafed to Isaiah at an early stage of his prophetic ministry. In the year that King Uzziah died, he saw "the Lord sitting upon a throne, high and lifted up, and his train filled the temple. Above it stood the seraphim; each one had six wings; with twain he covered his face, and with twain he covered his feet, and with twain he did fly. And one cried unto another and said, Holy, holy, holy is the Lord of Hosts; the whole earth is full of his glory. And the posts of the door moved at the voice of him that cried, and the house was filled with smoke." We cannot wonder that the prophet, confounded with that unearthly manifestation, was overwhelmed with dread, and exclaimed,

---

* On occasion of the annual Sermon, designed to refute the mediæval Jewish schools of prophetic interpretation.

"Woe is me, for I am undone, because I am a man of unclean lips, and I dwell in the midst of a people of unclean lips." But as he uttered his confession, one of the seraphims flew unto him, having a live coal in his hand, taken from the altar, and laid it upon his mouth, and said, "Lo, this hath touched thy lips, and thy iniquity is taken away and thy sin purged." And then he was warned to go and tell his nation, "Hear ye indeed, but understand not; and see ye indeed, but perceive not. Make the heart of this people fat, and make their ears heavy, and shut their eyes; lest they see with their eyes, and hear with their ears, and understand with their heart, and convert, and be healed."

There can be no doubt that it was a vision sent at once to cheer the prophet in his work, and to prepare him for it. He was to go to teach his countrymen; but Providence foresaw that they, with the inflexible tenacity of character which has ever been their marked national peculiarity, would refuse to listen to his message; and, therefore, he was given to feel that his ministrations, though they were not heeded on earth, were not unnoticed in heaven; that unclean as he was, and ministering to a people unclean, there was a seraph to fly to him with the assurance of mercy; and that though he might be led to think that the course of this world's history was for evil, yet the seraphim standing before the throne, and surveying things by the light of eternity, were chanting their song of triumph to the Holy, Holy, Holy, because they were permitted to witness that not the heavens only, but the earth also, was full of God's glory.

The history of Isaiah's ministrations to the Jewish people, has been repeated in every succeeding attempt made to bring them to a sense of their true condition. One messenger after another has been sent, and at last the Divine Son of God came forth. He came to His own, but His own received Him not. And the result has been that vengeance appears to have overtaken them; their vineyard has been taken away and given unto others.

If there were no other interest belonging to the Jewish nation than that which arises from the operation of merely ordinary laws in their history, they would yet be singled out as one of the most remarkable of peoples. The interest which belongs to them would indeed be unlike that which appertains to other nations. No mystery envelopes their origin, such as excites our curiosity with regard to many ancient races, which have left in their cities and their cemeteries, the traces of a civilization which must ever remain an enigma. No widely spread influences can be traced to them, such as those effects which Athenian cultivation has stamped indelibly on the world. No political example is offered in their history, of a people working out its liberties, and then imprinting its laws on a conquered world, such as gives to Roman history its enduring interest. Yet, in spite of the absence of these features, the peculiarity of their pertinacity of character, of their persecutions, and their continuance as a separate nation; scattered through every district of the civilized earth, yet not confounded with the masses of its population; strangers where they have long had a home; foreigners where they have long been naturalized; separated by an ineffaceable barrier from societies with which they hold the closest companionship;— these circumstances alone, if there were none of different and higher interest, would claim for their history and condition the attention of all who desire to understand the philosophy of man.

When, however, we superadd to this merely secular view

of their history the Divine aspect which revelation presents us of it, we feel that it stands out singly in the progress of the race. Other histories embody ideas; it is theirs alone which embodies *Divine* ideas. They stand out as the instruments of a special administration, and the possessors of a special religion. They appear as the rejectors of the Messiah whom they had long anticipated, and their dispersion is regarded as a Providential punishment for that act of ingratitude.

It is to the great fact of the rejection by the Jews of that Being whom we believe to have been the Messiah, that our attention is to be directed in the present discourse. It is not surprising that a philanthropic individual who felt a deep sympathy with those attempts which have been made to convert the Hebrew nation to Christianity, and a sincere interest in their welfare, should have desired that the subject should be brought before this University, and should accordingly have presented a gift to it a few years ago[*] for an annual Sermon, as he himself expressed it, "*on the application of the prophecies in the Holy Scriptures respecting the Messiah, to our Lord and Saviour Jesus Christ, with an especial view to the confuting the arguments of Jewish commentators, and the promoting the conversion to Christianity of the ancient people of God.*"

The subject of the present discourse is, therefore, defined by the wishes of that benefactor. I should have been glad if we could have investigated the various causes which have operated most forcibly in preventing the Jews from accepting the Christian faith. The brief space of our present service will, however, only permit of the review of a single one of them, the consideration of which will, I

* In 1848.

hope, nevertheless answer, in some humble manner, the noble purpose which our benefactor had at heart.

The cause to which I allude, is the fact that the Jews possess a literature directed against Christianity, which is not only taught to the mass of their nation, but is of sufficient subtilty and importance to command the respect, and in some sense satisfy the judgment, of their intellectual men. We are too apt to regard them as rejecting Christianity, simply because their fathers did so, and because they have never had the candor to reconsider the question. This, however, is not wholly the case. They possessed, especially in the middle ages, distinguished writers, who established a regular school of prophetical interpretation in answer to the Christian theory of the fulfilment of those prophecies which relate to the Messiah. And as the wish of the founder of this Sermon contemplated a reply to those writers, which are referred to in argument with the modern Jews, I think that it will not be a misemployment of your time if I first give such a brief sketch of Jewish theological literature as will enable you to understand the nature of their opposition to Christianity, more especially as I am not aware that any of those who have preached in past years on this subject have done so.*

---

* Two of the Sermons which had been preached in preceding years have been published; one by Dr. Marsh, in 1849, entitled, "Predicted History of the Messiah fulfilled in Jesus," and the other (in 1850) by Rev. C. Girdlestone, on "Messiah Pierced," both of which Sermons investigate a single passage of prophecy without presenting an introduction to the literature of the subject. The sources from which the facts for the present Sermon have been drawn are mostly enumerated in the notes. A useful work on the history of Jewish literature has been recently published by the Syriac scholar, Dr. Etheridge, entitled, "Jerusalem and Tiberias; Sora and Cordova; a Survey of the Religious and Scholastic Learning of the Jews."

Jewish literature has especially flourished at three different periods, and in three different lands;—in Judæa in the period which intervened between the return from captivity and the commencement of the Christian era; in Galilee and Mesopotamia from the 3d to the 8th century A.D.; and in Spain from the 10th to the 15th.

1. Few public events ever worked so mighty an effect on a nation in so short a period, as the captivity at Babylon wrought on the Jews.* It affected their social and intellectual life in modes which exist to the present day. It imparted to them a new language and a new written character;† it forever banished from them the practice of polygamy; it excited in them a lasting hatred of idolatry; it enlightened them on the doctrines of a future life and of moral duty, and by binding them in a common suffering and a common sorrow, extinguished for a time those unhappy feuds which had so often proved their ruin. But it was the effect which related to their literature with which we are now concerned. Their social state was so altered by their captivity that they were compelled to form an uninspired theological literature. For they returned home, as we have already hinted, with a new language. They had not only laid aside the old Hebrew alphabetical charac-

---

* See Milman's "Hist. of the Jews," b. ix.

† The view here intended is, that the Jews exchanged the old Hebrew or Samaritan character for the square Chaldee, which is now called Hebrew; the old character reappearing only in coins of the house of the Maccabees. The Hebrew language was also exchanged for Chaldee, or Eastern Aramaic; and in Galilee and the northern parts of Palestine, the pronunciation in later times probably approximated more to the Western Aramaic or Syriac. The original sources for forming a judgment on this question are given in Horne's "Introduction," vol. ii, ch. 1. See also Stuart's "Hebrew Grammar" (Introduction), and Marsh's "Lectures," part ii, pp. 136, *et seq*.

ter, but had adopted the forms of speech of the Eastern Aramaic, or, as it is commonly called, the Chaldee tongue. Hence their own law became unintelligible to them,* and the necessity for understanding it called forth a new order of interpreters, and a new literature of translations or paraphrases into the newly acquired tongue. These interpreters are known by the name of the Rabbins, and the translations by the name of Targums.

There were circumstances, too, in their social condition at that time, which gave increased importance to these writings and their authors. After the captivity, the Levites ceased to be the great instruments for teaching the people, and a new order of teachers arose in those separate little centres of worship, which grew up under the name of synagogues.† It is not difficult to see how such an order would gradually gain power. Parallels are offered to us in other countries, as, for example, in the epoch when the Latin language ceased to be spoken, and was changed into the various tongues of modern Europe; or in India, in the age when the Sanscrit ceased to be the vernacular tongue, and yet continued to be the depository of the religious creed. In both epochs alike there continued to be a learned language in the hands of an educated order; and this order naturally acquired intellectual and afterwards spiritual influence; in the one case there arose the Roman Catholic priesthood, and in the other the Brahminical; in both they first became the translators, and afterwards the interpreters of the ancient religious books.

It was this department of interpretation which gave to the Rabbins the opportunity of insinuating into the Jewish

---

* Compare Neh. 8 : 8.
† See Milman's "Hist. of the Jews," iii, book 18.

mind the body of traditional doctrine distinct from the written word of God, which formed the strength of Pharisaism, against which our Lord so often levelled His addresses.* It does not, however, appear that this new system of doctrine was at that early period committed to writing, or that the paraphrastic translations or Targums of the ancient Scriptures, of which we possess the copies, were composed till near the end of the first of those periods into which we divided Jewish theological literature.† Nor would the notice of them have fallen properly within our province, if it were not on account of the importance which they assume in controversy, as recording the interpretations assigned at that time to certain passages in the Old Testament.

Independently, however, of this consideration, the time will not be lost that has been spent in thus viewing the morning of Jewish uninspired literature. The productions of that age may be few, but in noticing the causes which thus created a literature, we have ascended to the fountain head of the waters which ultimately expand themselves into broad streams. And as the student who wishes to understand the history of Art busies himself with the study of the age when it was struggling to emancipate itself from the crudeness which cramped its early efforts towards a free development, so, if we would view the full daylight of a nation's literature, we must watch its sun rising amid twilight, and battling with the mists which obscured its early brightness.

2. An interval of two centuries separates the second age of the national literature from the first,—an interval during

---

* *E. g.* in Mark 7.
† See Bartolocci, "Bibliothec. Magn. Rabbin," vol. iii.

which the nation, after struggling for its independence, and after manifesting a heroic patriotism, even in the hour of its deepest gloom, had been finally removed from that city which, for more than twelve hundred years, had been the metropolis of the religion and the race. Yet in spite of their exile, the Jews were able to maintain their nationality, and to form centres of Jewish life in different spots in heathen lands. Two places were selected by them as their especial homes. The one was Galilee, which was under the Byzantine power; the other was Mesopotamia, under the Sassanian dynasty of Persia. In each a centre of government existed for the dispersed Jewish people, whence they received their creed, and to which they yielded spiritual obedience. In each resided a Patriarch who regulated their whole system of education, and directed, by means of legates, the affairs of his nation in other lands, exercising a power which offers no unapt analogy in miniature to the combined spiritual and temporal powers afterwards exercised by the Popes of Rome, or by the Mahometan Caliphs.

While these Patriarchs flourished, their abodes—the one at Tiberias, the other in Babylonia—were the places around which were gathered schools of the most learned Jews, and to which the youth of their people, scattered in other lands, betook themselves to receive their education. The literature taught was entirely theological; but a regular and well-ordered school of it existed, which has produced works which form the standard national literature, even at this day. The theological studies embraced the two subjects which we are accustomed to call Biblical Criticism and Biblical Interpretation. The books on the former subject related to the determination of the genuine text of the ancient Scriptures, and are called the Masora; the latter related to the meaning of the text, and are called the

Talmud. It is to the school of Tiberias that we owe the system of Biblical criticism. It was the teachers gathered there about the year A.D. 400, who collated manuscripts, and determined and arranged the text, performing much the same kind of office which the Christian critic Origen had so honorably executed at Alexandria two centuries earlier; and it was probably at that time that they attempted to fix the ancient pronunciation of the Hebrew by the invention of the vowel points,* with which that language is now usually written.

While the school of criticism was restricted to Tiberias, that of interpretation was even more cultivated by the Jews of Babylonia. A systematic digest was there made of the traditional interpretations which had grown up through centuries. It was named the Talmud, and contained two parts, the Mishna, or text of the traditions, and the Gemara, or commentary on them. The difference will be understood by those who are familiar with the legislation of Justinian.† The Mishna was like his Code, embodying the national laws; the Gemara, like his Pandects, embodying the mass of precedents. This system of interpretation is received by most Jews with the same reverence which they attach to the Scriptures. It embodies, according to their belief, an oral tradition originally revealed from heaven, and handed down co-ordinately with the sacred volume. Nor ought it to escape our notice how closely herein their feeling resembles that with which the Roman

---

* The date of the introduction of the vowel points was a subject much debated among the great Hebrew scholars of the 17th century. The references for investigating it are to be found in Horne's " Introduction," vol. ii, ch. 1, sect. 1 ; and Marsh's " Lect.," part ii. The opinion of Cappel here adopted is now generally received.

† Gibbon's " Decline and Fall," ch. 44.

Catholic regards the teaching of the tradition which rests on the authority of the Church;—a circumstance attributable in its origin, as we have already stated, to the fact of the religious teaching resting with a learned order at a time when the majority of the people were unable to investigate for themselves; and a clear example how remarkably the various events of the middle ages, in which there seems at first sight no regularity, are really reducible to the same causes, and capable of being generalized into the same laws; being but manifestations of the similar state of society which existed in different countries which were at the same stage of political growth.

It is this reverent regard which the Jew bears to the Talmud that renders it of importance in controversy. It stands to him as the Bible does to the Protestant; or as the Vedas to the Brahmin; or as the decrees of the Church to the Roman Catholic. But there is also another value in it, viz., that in spite of the mass of allegorical and fanciful interpretation which it contains, it conveys the first example of the unreal, and, as we believe, forced interpretations which the Jews began to find it necessary to impose on the old prophecies, in order to wrest them from the use to which the Christian writers applied them.

3. We shall now proceed to sketch the third period, or, as it may be truly called, the golden age of Jewish literature, which existed in Spain from the 10th to the 15th century.

The state of the Hebrew nation in this period forms nearly the only bright spot in the sad picture of their history.

It cannot be a subject for surprise, that when the Mahometan conquerors, at the beginning of the 8th century, crossed the straits which separate Africa from Europe to conquer the Spanish peninsula, they were hailed by the Jews—who had been bowed down under the oppression of the Visigoths—as friends and deliverers. Under the generous protection of this race of conquerors, the Jews lived in happiness and increased in material prosperity, maintaining a commerce between the eastern and western parts of that sea, which, on three of its shores, was inclosed by the vast Mahometan empire; and it ought to be an instructive lesson to consider that it was under the shelter of the followers of the false Prophet that they found the protection which they sought in vain from the followers of Him, whose very last prayer had been for their race: "Father, forgive them, for they know not what they do."

It was the safety and wealth that the nation possessed, which enabled its superior spirits to devote themselves to intellectual pursuits. That may truly be said to have been the noonday of Jewish literature. Not only in theology, but in the art of poetry, and in science, there arose distinguished writers. At a time when the rest of Europe was enshrouded in darkness, broken only by the little lamp of knowledge which had been borrowed from the Moors, science and learning were beginning to shed their rays over the Mahometan kingdom of the peninsula. Algebra and the abstract sciences were eagerly pursued by them, and Jewish astronomers were employed in constructing the Alphonsine tables, the interest of which is well known.* Discoveries were made in anatomical science by Jews, and the chief physicians in Europe were taken from that nation; whilst others of them, through their knowledge of banking and finance, rose to high ministerial functions and offices in the courts of the caliphs. Possessed of equal rights with their Mahometan fellow-subjects, it seemed as

---

* The tables of Alphonso X, in 1252. (De Castro's Hist., p. 62.)

if they forgot that they were in a strange land; their harp no longer hung silent upon the willows; the spirit of their ancient psalmody revived; and many of their lyrics remain, cramped indeed by the unnatural adoption of the Arabian metre, and by the use of rhyme, yet breathing in their matter and meaning that spirit of poetic inspiration which always commands the sympathies and awakens the response of the general heart.*

It was in the department of theology, however, that their literature was most distinguished. It was an accident, which about the middle of the 8th century brought a distinguished Jewish theologian from the East into Spain.† Welcomed by his countrymen, he immediately opened a school of Jewish literature in the then rising University of Cordova. The higher Jews, nevertheless, still continued to send their children to the schools of Egypt or Babylon to receive their education, until the schism in the caliphate and the persecutions commenced in the 11th century by the Egyptian (Fatimite) dynasty drove the Jewish professors to seek refuge under the enlightened sway of the Spanish caliphs. The effect of this immigration on the revival of Jewish learning was almost as marked as that which was seen in the analogous case of the revival of Greek literature in the 15th century, when the Greek population of Constantinople retreated into Western Europe on the taking of their city by the Mahometans.

Schools of Jewish theology not only sprung up immediately in Cordova and the other great cities, such as Seville and Granada, which formed the glory of the south of Spain, but their influence extended across the chain of

---

* The Mediæval Jewish Poetry will be found translated into German, in "Die Synogale Poesie des Mittelalters, von Dr. Zunz; Berlin, 1855. See also his work, "Zur Geschichte und Literatur; Berlin, 1845."

† Named Moses. See Milman's "Hist. of Jews," iii, 285.

mountains which hems in the province of Andalusia, and made itself felt in the Universities of Toledo and Valencia, and even as far north as Barcelona. The great subjects of study in these schools were the principles of Biblical, and especially of prophetical interpretation. Though much tied down by the authority of the Talmud, the teachers still felt that its system of interpretation was often fanciful, its great fault being that it partook of the common property of Oriental thought, of assigning an allegorical meaning to that which is literal and fact.* Rejecting, therefore, an allegorical interpretation, they adopted a literal and grammatical one, and accordingly laid a basis for it in the careful study of the structure and genius of their own tongue.

Thus far their principles may have been sincere, and suggested by an honest perception of the improprieties of the ordinary system of interpretation; yet it must be added that the chief motive, which is at once apparent in many of their interpretations of particular passages, is the design of giving such a meaning to them as to destroy the force of the Christian interpretation of them.

It may be well, for the sake of giving individuality to these writers, to name the principal of them.† The early part of the 12th century produced three, viz., Jarchi, surnamed Rashi, Aben Ezra, and David Kimchi; and the

* This allegorical mode of interpretation became common about the time of Philo (A.D. 40). See Essay on Philo, in Prof. Jowett's work on St. Paul's Epistles.

† For information on these and other subjects, see J. B. De Rossi's "Dizionario Storico degli Autori Ebrei e dello loro opere;" also, " Hist. of Jews of Spain and Portugal, by E. H. Lindo, 1848;" also, " Hist. of the Jews in Spain, by Don Adolfo de Castro, 1851;" also the works of Calmet, Basnage, Gaffarelli, Delitzsch, Julius Fürst, and Jo. Chr. Wolf.

latter part of it produced one, viz., Maimonides. The first of these, Jarchi, was not, strictly speaking, a Spanish Jew. He lived in the northeast of France, and was the cotemporary and disciple of Abelard, and of other distinguished men, who in that age adorned the University of Paris. His mind was cultivated by extensive travel, and his commentaries are creditable to the judgment of their writer. The second of them, Aben Ezra, taught a few years later at Cordova. Eminent in his own day for his general cultivation, and for his acquaintance with foreign tongues, he is now known only as the author of a subtle commentary, which will bear comparison with those of a better age. The third, Kimchi, taught also in Spain, and is allowed by both Jews and Christians alike, to be a commentator remarkable for power of language, profoundness of knowledge, and clearness of method. The other name which we enumerated, is more generally known, viz., Maimonides. Educated at Cordova by Averroes, the celebrated commentator on Aristotle, he was thoroughly acquainted with Greek philosophy, as well as with Jewish theology. Accordingly he rose above the level of a mere commentator. Under the garb of a theologian he was really a philosopher. His purpose formed no unworthy parallel to that of the Christian Aquinas, who lived in Italy about half a century later. Just as that great thinker aimed at giving a universal philosophy, which, grasping in one magnificent generalization the worlds of matter and of mind, might assign to the Christian and ecclesiastical doctrine its true position in such a scheme; so Maimonides, endeavored to evolve a universal philosophy, from which the Rabinical conceptions of the Talmud might be natural corollaries. Both were trammelled by a body of doctrine which they neither desired nor were able to reject. But

Maimonides, less fortunate than Aquinas, was deemed to have trespassed on the received creed; and not only were his opinions the means of producing theological feuds, but their author was compelled to quit Spain, and die an exile in a foreign land.

After Maimonides the glory of the Jewish people began to decline. Their literature became, indeed, known in foreign lands, and the system of Pantheistic philosophy called the Cabbala, was even reproduced with approval in Florence, in the brightest period of Italian literature.* But in Spain, the events which followed tended to extinguish their literature. The tide of Christian conquest, which had steadily set in from the North, overflowed, about the year A.D. 1240, the valley of the Guadalquiver, and Cordova and Seville were retaken by the Christian kings of Spain. And though the conqueror transferred the Jewish teachers to Toledo, and offered them his protection, succeeding sovereigns persecuted them, and one only name stands out in their theological literature as a writer against Christianity, viz., Abarbanel, whose name is well known in connection with the court of Ferdinand and Isabella, and the circumstances of the expulsion of the Jews from Spain. It is necessary, however, to add to this list one more name, that of an individual who flourished in Lithuania about the close of the 16th century. It is the Rabbin Isaac, the author of the most complete defence of the Jewish creed, and the most subtle and controversial attack on our religion which has ever been written, and which, along with those previously noticed, is the standard authority with the modern Jews.

The history which we have now completed will, I trust, not have proved uninteresting, as it assuredly is not irrele-

---

* By Pico di Mirandola. See Hallam's "Hist. of Lit.." i, 3, 202.

vant to the subject of our present sermon, which is the establishment of the Christian interpretation of prophecies relating to the Messiah, against the views of the Jewish commentators.

We shall proceed accordingly to notice the proof of the Messiahship of our Lord Jesus Christ, drawn from the prophecies of the Old Testament, with a special reference to the refutation of the Spanish school of Jewish interpretation.

The controversy between the Jew and the Christian consists in this. Both admit the existence of a body of ancient prophecy predicting a Messiah, but they differ in their interpretation of it. The one asserts that the Messiah has not yet come, the other claims for Jesus of Nazareth the fulfilment of those anticipations. How shall this controversy be decided?*

One of the most simple methods would be this: Let us imagine ourselves to have been living at the period of the utterance of these prophecies, and endeavoring to collect from them the conceptions of this future personage which they would have been likely to supply. What idea should we have formed to ourselves of him? We should have fixed the date of his appearing before the power of the Jewish race should depart; for the sceptre was not to depart till Shiloh came.† We should have placed it more exactly, as Daniel tells us, within seventy weeks, *i. e.*, four hundred and ninety years, from the edict of Artaxerxes for the rebuilding of Jerusalem.‡ We should have expected

---

* The following arguments are partly condensed from one of Dr. McCaul's "Warburton Lectures on the Prophecies" (1846), and partly suggested by a periodical formerly published by him, entitled, "The Old Paths, or a Comparison of the Principles and Doctrines of Modern Judaism with the Religion of Moses and the Prophets," 8vo. 1837.

† Gen. 49: 10.   ‡ Dan. 9: 25.

the continuance of the second temple until the appearance of the Messiah, for Haggai declared that the desired of all nations should come in it.* We should have anticipated some predecessor to come in the spirit of Elias to prognosticate his approach.† The prophecy of Micah would not have left us ignorant as to the place where we might expect his appearance, for it was to be in the despised Bethlehem Ephratah.‡ Extraordinary combinations of qualities were to be expected in him. A virgin of the lineage of David was to conceive and bear him;§ and yet he was to be in some mysterious manner the Son of God, whose goings forth have been from of old, from everlasting.|| He was to exercise the office of a prophet, and to imitate the ancient Moses;¶ his prophetic mission was to commence in Galilee, for it was said that there the people should see a great light.** The wonders of the old prophets were to be reproduced in him; the eyes of the blind to be opened, and the ears of the deaf to be unstopped.†† He was to be a priest too of a new kind.‡‡ The government also should be upon his shoulder, and his name should be called Wonderful, Counsellor, the Mighty God, the Everlasting Father, the Prince of Peace.§§ Kings should fall down before him.|||| Of the increase of his government there was to be no end.¶¶ Yet along with this greatness how should we have reconciled the other qualities of which we read? He was to be despised and rejected of men, a man of deep sorrow, despised by his own friends, acquainted with griefs;*** how

* Hagg. 2:7.
† Mal. 3:1; 4:5.
‡ Micah 5:2.
§ Isa. 7:14.
|| Micah 5:2.
¶ Deut. 18:15.
** Isa. 9:1.
†† Isa. 35:5.
‡‡ Ps. 110:4.
§§ Isa. 9:6.
|||| Ps. 72:11.
¶¶ Isa. 9:7.
*** Isa. 53:3.

should we have harmonized his universal reign with the lowly riding on an ass,* and with his being weighed against thirty pieces of silver?† Lastly, he was to be cut off, but not for himself. Cruelly mocked, he was to die with the wicked,‡ his garments were to be parted,§ and yet he was to escape the fate of ordinary malefactors; his bones were not to be broken, though his side was to be pierced, and he was to make his grave with the rich.‖ But his soul was not to be left in the grave;¶ and he was to receive gifts for men, and lead captivity captive,** and pour out his Spirit, and sons and daughters were to prophesy.††

Such is the conception that might have been formed of the future Messiah, when the roll of the Old Testament prophecy was closed, and the voice of Providence sealed up the words of the prophecy of that book. And can a candid mind doubt whether the being has yet appeared who answers to this anticipation? Who is it that is yonder in the town of Bethlehem,—the young infant on the knee of his virgin mother? Who is it that is near the Jordan, saluted as the lamb of God by the hermit prophet, at whose wondrous teaching the Jews have repented as of old at that of Elias on Mount Carmel? Who is it that toils day by day among the highlands of Galilee, never rejecting the prayer of the needy, dispensing his mercy? Who is it that kneels yonder, lonely in the garden, bathed with the dews of night, bleeding with the agony of a soul exceeding sorrowful, muttering, as if in the consciousness that he was not cut off for his own sin, "Father, if it be possible, let this cup pass from me?" What has caused Nature to

---

* Zech. 9: 9. † Zech. 11: 12, 13.
‡ Isa. 53: 9. § P's. 22: 18.
‖ Isa. 53: 9. ¶ P's. 16: 18.
** P's. 68: 18. †† Joel 2: 28.

respond by a miraculous darkness as one of you malefactors has exclaimed, "It is finished," and has given up the ghost? Or, where can his buried body be gone? "for some one has taken him away," exclaims his weeping follower, Magdalen, "and we know not where they have laid him." Tell me, I ask you, if it is not the individual, God and man, Saviour and sufferer, Prophet and sacrifice, of whom the prophets wrote and spoke? Tell me which is history and which prophecy, the statements made hundreds of years before his appearance, or the simple unadorned narrative of His loving disciples?

So close a coincidence betwixt prophecy and history carries a moral force which is well-nigh irresistible. Whatever difficulties may attend the interpretation of prophecy, whatever discredit the rash haste of undisciplined minds may in this day have cast upon it, whatever suspicion the modern investigations on the nature of evidence may have thrown upon analogical reasoning,* such an accumulative proof as this, is powerful enough to outweigh them. If the number of the coincidences were small, or their application merely general, we might doubt whether the interpretation of the prophecies was not fanciful; but their multitude, their minuteness, and their variety, forbid the possibility. The value of collective analogies like those with which we deal in assigning the meaning of prophecies, depends upon the principle which is commonly called "circumstantial evidence." As they increase in number, in intricacy, in variety, the improbability of a chance coincidence becomes immensely heightened. The hemp threads which compose a coil of rope are separately weak, yet when united and intertwined, they form the tenacious cable which has

---

* See Mill's "Logic," vol. ii, ch. 20.

strength enough to resist the force of pressure, or to allow the ships to ride at their anchorage in safety as they rise and fall before the heavy swell of the rolling tide.*

But it will naturally occur to any of us to whom the proof of the Messiahship of Jesus appears so clear, to demand the grounds on which the Jewish writers reject it. Their reasons are principally three, which we now proceed to consider,† viz. :—

(1.) The historical one, that the Sanhedrim and Jewish authorities, in our Lord's lifetime, who had every means of examining the claims of Jesus, and who were actually predisposed to accept a Messiah about that time, rejected His claims.

(2.) The philosophical one, that an incarnation of a Divine Being is an impossibility in the nature of things, as well as contrary to the analogy of the Mosaic dispensation.

(3.) The critical one, that the prophecies which relate to the Messiah are in great part either mistranslated or misinterpreted. The first of these is the popular objection; the second, that of Maimonides; the third, that of the other members of the school of Spanish commentators.

---

* Circumstantial evidence is logically invalid, in consequence of the technical fault of the middle term being undistributed in each of the syllogisms which compose it. Each syllogism is the ἀνώνυμον σημεῖον of Aristotle's "Rhet." (i, ch. 2), and falls into the second figure. But though each argument is separately weak, the convergence of a large number, in proof of the same conclusion, possesses by the doctrine of chances a high logical probability, and in its moral effect is irresistible. It is this which constitutes much of the strength of Butler's "Analogy." The single analogies are weak, but the number and convergence of them towards the same point have the force of strong circumstantial evidence.

† These three reasons are well discussed in McCaul's "Warburton Lectures," to which reference has already been made.

1. In reply to the first of these arguments, it is sufficient to urge that we are not in doubt as to the character of the Sanhedrim in the time of our Saviour. Josephus remains as an unprejudiced witness of the profligacy, the corruption, the worldly and sceptical spirit of that body. It is natural that the Sadducean or sceptical party in that assembly should have rejected our Lord's pretensions; while, with regard to the Pharisaic section, any one who will refer to the Talmud, which embodies the traditional theology of Pharisaism, will feel convinced that those whose minds were enslaved by such puerilities, and whose faith in a Messiah, firm though it may have been, was in the appearance of an earthly sovereign, were not capable of being correct judges of the claims of One who preached a doctrine which was opposed to theirs, and whose life ran counter to their preconceptions. This reply, it will be observed, is founded on Jewish authorities; but if the Christian evangelists be further appealed to as cotemporary, though (let us for the moment admit) one-sided, witnesses, a confirmation of this view is attained; for from them we learn that no proper or candid investigations of our Lord's claims were ever made.

2. In passing to the second objection, which is urged on the part of the Jews, we no longer encounter one that is merely superficial, but one which comes with higher pretensions, and is supported by a great name. It is Maimonides who urges the impossibility of the incarnation of deity and the contrariety of such an idea to the Mosaic economy.

We have before shown that the philosophy of Maimonides was partly founded on that of Aristotle; but to this ingredient was also added a considerable admixture of the Oriental philosophy of Zoroaster. It was this system which, learned by the Jews during the captivity, and wrought into a system called the Cabbala, and embodied in a work of the early ages, called the Zohar, infected more or less their men of superior minds to the latest period of the middle ages. Without inquiring what it was in itself, it would have amounted, as realized by Maimonides, to a system of pantheism, if it had not been modified by his Jewish education. If you can combine the Jewish idea of one personal God with the pantheistic notion of the impossibility of his attributes being separated from the universe, you will understand Maimonides's idea of the Divine Being. It will not be hard for those who comprehend the subject to perceive how such a person would feel an aversion to the idea of the incarnation and suffering of divinity. And we may be excused from pausing to refute a view which receives its answer by the refutation of the theory from which it is a corollary.

The other part of his objection—viz., that the idea of an incarnate Messiah was contrary to God's revelation to the Jews—was founded on his view of the purpose of the law of Moses. He has left us a work on this subject,* wherein he endeavors to show that the central thought of the ancient dispensation was to lead the Jews to the knowledge of God and the abandonment of idolatry. Hence he plausibly considers that the idea of Jesus being the Messiah would be a glaring instance of the very infringement of the command which forbade the making any similitude of the Almighty, for the violation of which the Jews had so often suffered, and the obedience to which it had been the great purpose of the Jewish economy to establish.

---

* "The Reasons of the Laws of Moses, from the *More Nevochim*," translated by Dr. Townley.

The answer to this objection is to be found, first, in the historical statements of the repeated appearances of the Divine Being under human form in both the Patriarchal and Jewish times; and, secondly, in denying that the worship of God and man in one Christ is obnoxious to the charge of producing that moral evil on the mind which the old forbidden idolatry confessedly effected.

3. Leaving these objections urged by the Jews against Christianity, we pass to the third,—the assertion that the prophecies supposed to apply to the Messiah are either mistranslated or misinterpreted. With regard to the former charge, it is not necessary to make any observations, because the defence of it is now given up by their own writers, and because the ancient versions of which we spoke, called the Targums, frequently support the Christian translation of the disputed texts. Nor shall I say anything in reference to the subject of types, though it truly belongs to this point; for type is but a prediction by *action*, as prophecy is by *words*; and the Jews are unable to find any solution so plausible to account for the sacrifices of their own law as that which is offered by the writer of the Epistle to the Hebrews. If Jewish sacrifice was (as they believe) from heaven, it is an enigma insoluble except by Christianity. We restrict ourselves accordingly to the charge urged by the Jews of misinterpretation of prophecy.

The prophetic texts which are made the grounds of dispute are treated by the Hebrew school of commentators in two different modes. The one class of passages is explained by giving them a local and literal sense, as, for example, applying the passage of Isaiah, " Unto us a child is born, and he shall be called wonderful,"* to Hezekiah; the other class is where apparently contradictory attributes are applied to the Messiah, as when he is described as "sitting on the throne of David to order it and to establish it forever;"* and in another place as "wounded for sin and bruised for transgression, and making his grave with the wicked."† This class is explained by supposing that there were to be two Messiahs,—one to suffer, the son of Joseph; the other to reign, the son of David.

The majority of the passages in dispute are contained, as would be supposed, in the Psalms, in Isaiah, and in Zechariah; these being, for this reason, called emphatically the three evangelical books of the Old Testament.

We may adduce as an example of the passages which are explained to refer to some other person than the Messiah, the magnificent text where Zechariah breaks out into the strain,‡ "Awake, O sword, against my shepherd, and against the man that is my fellow, saith the Lord of Hosts: smite the shepherd, and the sheep shall be scattered: and I will turn mine hand upon the little ones." The commentator, Kimchi,§ and others to whom we have alluded, assign this text to various heathen kings, and understand them to be described in it ironically as Jehovah's fellows. What is the line which might be adopted in reply to this view? It is, first, that the Targums and the Talmud both apply it to the Messiah, thereby proving that such was the view of its meaning adopted by the ancient Jewish Church; and, secondly, that the new interpretation, if even it will fit the

---

* Isa. 9: 7.   † Isa. 53: 5–9.   ‡ Zech. 13: 7.

§ The Commentary of Kimchi on Zechariah has been translated, with a useful Commentary, by Dr. McCaul; to which, with the other works on the Jews by the same writer, such as the intellectual state of Rabbinical Jews, in ch. 1 of "Sketches of Judaism and the Jews," the author of this Sermon is under large obligations.

passage under consideration, would not suit the context, because the same person who is here called "Jehovah's fellow" and "the shepherd of the people" is predicted to be sold for thirty pieces of silver, to be abhorred by the rich, to be loved by the poor, and to be cut off before the scattering of the Jews. Here are several circumstances which must be combined in any theory of the meaning of this passage. For the laws of critical interpretation must be amenable to the tests which regulate hypotheses in the sciences; and if we are accustomed to hold it to be the highest confirmation of a scientific theory that it is adequate to explain the various phenomena to which it is applied,* is it too much to require that such a test shall be regarded as equally decisive in the science of Scripture hermeneutics?

The other class of texts to which we alluded as describing at once the glory and the humiliation of the Messiah is too well known and too numerous to require quotation. The pathetic description† of the person "smitten and stricken," whose "soul was made an offering for sin," "upon whom was our peace, and upon whom God laid the iniquity of us all," is a sufficient example, especially as the whole body of Jewish mediæval commentators admit (as their fathers did) that it applies to the Messiah. But they explain it to apply to a different Messiah from those texts which describe a Messiah who is to reign. In reply to this view, it is sufficient to state that we can prove historically that it was unknown until it was invented, in the process of con-

---

* The two tests of scientific hypothesis usually given are, that the supposed cause be *vera*, and that it be *adæquata*. See the interpretation offered of them in Mill's "Logic," vol. ii, b. iii, ch. 14.

† Isaiah 53.

troversy, for the purpose of refuting Christianity; and that not only is there no example of a promise of two Messiahs, but such a view is contrary to passages in the prophets, where the same person is spoken of in the same verse under the two capacities of monarch and sufferer, triumphant and abased.

We have now sketched the Jewish objections and the mode of their refutation; and we might, if time would allow, accumulate direct arguments in favor of the Christian view of the advent of the Messiah in the person of Christ. But it may be permitted us to remark that, in any attempt to draw inferences from the prophecies of the Bible, we encounter a difficulty, arising from the want of any fixed principles of interpretation. In the explanation of the other mystical parts of the Bible, such as the type, the allegory, or the parable, theological science has to a great extent ascertained fixed methods of interpretation; but in the explanation of the prophecies there is no such rule. We have hinted already what must be the plan for discovering such rules. It must be by a careful study (1) of the nature and value of analogical evidence; (2) of the nature and meaning of Hebrew symbolical language; and (3) by the analysis of some few instances in which prophecies have been undoubtedly fulfilled, in order that rules derived from the two former methods may be tested and verified by the latter. Such a system of prophetical interpretation is yet to be created. Yet it will not, I hope, be improper in this place if I adduce the name of one individual—an ornament of our University and Church, who, had he lived longer, might perhaps have given such a system, and who has left a work, cautious, logical, original, and philosophical, which is earnestly to be recommended to every theological student. I allude to Mr.

Davison's "Lectures on the Structure, Use, and Interpretation of Prophecy." It is one of those few works which can be pointed to as assisting in raising theology to the dignity of a fixed science.

In conclusion, I wish to draw two brief inferences; the first of which concerns our duty to the Jews, the second our duty to ourselves.

1. We have sketched only one great feature of modern Jewish life, but it is one which will commend itself above all others to the sympathies of this present congregation. We have, nevertheless, seen enough to serve as an argument to our consciences in claiming for the Jewish nation our interest and our respect. Whatever may be the other hindrances to such a conversion of them as shall admit them to share the blessings of Christianity, the obstacle, at least, which arises from intellectual prejudice may be, we may hope, subdued by argument, or dissipated by kindness. It was with a purpose of this kind that a few years ago a Bishopric was founded at Jerusalem, and that missionary efforts have been carried on among the Jews abroad, and a missionary colony of converted Jews established at home. Yet it is not in great efforts like these, but in smaller acts of kindliness, whensoever they come in contact with our civilization, that we may attempt to prove our religion by our acts. In England, at least, we have done a great deal in this direction by admitting them to social and municipal rights. For, verily, we are great debtors to them for the shameful persecutions which Christians have exercised towards them in the middle ages. History records no massacre more ruthless than that slaughter of the Jews which was committed on the Rhine by the hordes of savages who went forth under the banner of the Cross to fight in the first crusade; and there

is no one here who has not read the sad story of the expulsion of the Hebrew people from the Spanish peninsula.* That is a thrilling narrative in history, the scene of which is laid on the banks of the Loire, which presents to our imagination the picture of a whole people, with their wives and little ones, crossing the broad stream to escape the tyranny which they dreaded;† but there is something still more sad, and that touches the human sympathies with a keener sense of shame, in the sight of yon people landed on the seashore of Morocco, homeless, tentless, starving, driven forth from the towns which had been their homes, to wander over the earth, until death should release them from their woe. There is something in the emigration of the heroic people of La Vendée which we can look upon with admiration, for they are themselves the authors of the stern resolve to forsake forever the fastnesses where they can be no longer free; and they march, cheered with the hope of obtaining the protection of that flag, which, float where it may, marks out the home of the free and the refuge for the oppressed. But, oh! there is not one ray of light or of hope to illuminate the dark scene of the expulsion of the Jews from the garden-like valleys of the Peninsula, which had been associated with the golden age of their modern history, when they were driven forth against their will, with their honest industry snatched from them, expelled for no offence, the victims of priestly bigotry, wanderers without a friend or a shelter. Think you not that, as the thousands of them yielded up their lives in that sad emigration, under the force of hunger, of heat,

---

* The narratives may be found in Milman's "Hist. of Jews."

† The migration of the people of La Vendée. See Alison's "Europe" (first series), ch. 12.

or of toil, there ascended with their parting breath into the ear of a God of mercy, a cry of vengeance against the land and the people that had sent them forth? Think you not that every effort of those who profess the faith of those persecutors to wipe out and atone by kindness for the cruelty of those who called themselves by the name of Christ, must be well-pleasing to that Being, who spent His life in sympathy, whose last prayer was for His enemies, and who, though enthroned within the Shechinah of eternal glory, is still, in all the strength of human sympathies, touched with the feeling of human infirmity?

We cannot penetrate the darkness which overhangs the coming history of God's ancient people; we cannot venture to predict that our efforts to impart to them our civilization and our religion shall be successful; we can, however, rest certain that it is our duty to endeavor to make them participate in these blessings. Yet the eye of hope, if it reads truly the unfulfilled prophecies which affect their race, cannot but think that the Jews are still reserved for a glorious destiny. It would seem that by some mighty impulse, and at some mysterious signal, their scattered tribes shall arise from the mountains, and valleys, and islands of the earth, and hasten to recognize the long-expected Christ. Yes! their Messiah shall one day come to them, but not in the clouds of heaven. With the still small voice of conscience and of his spirit, He shall manifest himself to their souls. Each of them shall see, as it were, the vision which Isaiah saw, "the Lord high and lifted up," the radiant form of Jesus throned in the fire and cloud, attended by the song of the seraphim, revealed to the eye of the soul; and, looking by faith on him whom they have pierced, they shall recognize in him their long-expected Christ. And as their hearts sink within them at the thought, and as each exclaims, " Woe is me, for I am undone; for I am a man of unclean lips," the seraph shall be sent forth with a live coal to declare that their iniquity is pardoned, and that their sin is covered.

2. Finally, our subject is not without a lesson to ourselves. For though our earthly mission may not be towards the Jews, each of us has a duty to perform in the world, and the vision of Isaiah opens up to us the spirit in which alone we can seek to perform it rightly.

There is no spot on earth where a larger number of men of noble hopes or of high principle are gathered than in this University. And they who make it their business to gain the confidence of those whom they are privileged to instruct, well know that in the hearts of many students there dwells a deep and earnest wish to make their life here the means of preparation for a life of usefulness hereafter. Before many years are past, each one of us must go forth into the world to influence it or to be influenced by it. It will then lie in the power of each one to do something, however little, for God and for goodness. Amid the squalid thousands of our crowded towns, or in the retirement of the rural chapelry; amid the infection of hospitals, or bending over the bed of poverty; amid the scenes of ordinary life, and in acts of common philanthropy, we may seek to work the work of Christ. But if we would be the means of doing so, we must not take our religious tone from the world, but must introduce into society some ingredient of goodness which it does not possess. That ingredient comes down from heaven. It is the power of God's Spirit which alone can give it us. It is He alone who can kindle in our souls the flame of love which shall burn with inextinguishable glory for His honor and man's welfare. And the way to obtain that Divine help is the same as in the case of Isaiah of old. We must contrast our unworthiness with the Divine purity, and learn to drop the tear of penitence, and pour out day by day, from our inmost hearts, the cry: " Woe is me, for I am undone; for I am a man of unclean lips, and I dwell in the midst of a people of unclean lips." And as soon as we shall have done this, the seraphim will be commissioned to take the live coal from the altar of incense, and to touch our lips. It is from the altar that the seraph brings the coal. It is not for our sake merely that God is merciful, but because there is an altar of incense in His presence; and our prayers, mixed with that incense of our Saviour's intercession, rise up as a memorial before God. Unless we catch Isaiah's spirit we cannot be prepared for the prophetic work. It is only when the seraph has touched our lips, and our sins are cleansed, that we can hope to receive the preparation which shall fit us for our ministrations of love.

And in our life of labor let us ever keep before us the sense of our unworthiness and of God's mercy to us; and then, when life draws to a close, if we stand trembling at the thought of labors apparently useless, and lament in the words, " Woe is me, for I am an unclean man," the angel shall be again commissioned with the symbol of mercy to cleanse our sins; and our purified souls shall be admitted to see the Lord high and lifted up, eye to eye, spirit to spirit, and to join in the seraph song of " Holy, Holy, Holy!"

NOTE.

Pp. 124 *et seq.* The integrity of the writings of Isaiah, Zechariah, and the other prophets is here assumed without discussion, as it is not one of the questions in dispute with the Jewish interpreters. The Christology of the Old Testament, to which allusion is made in this Sermon, may be studied in Hengstenberg's work on the subject.

SERMON V.

THE DOCTRINE OF THE HOLY TRINITY.

(PREACHED BEFORE THE UNIVERSITY, JUNE 20, 1858.)

EPHESIANS 2 : 18.

"For through Him we both have access by one Spirit unto the Father."

THE doctrine of the Holy Trinity, which the Apostle implies in these words, is the centre of a group of Christian doctrines which may fairly be said not to have been explicitly known antecedently to the teachings of our Saviour and his Apostles. More than even other doctrines, this had hardly been guessed at by heathen speculation, hardly understood by Jewish inspiration. It stands in majestic isolation from other truths, a vision of God incomprehensible, the mystery of mysteries. We can find analogies and explanations of other doctrines in the world of nature, physical or moral, but of this we can discover none.* The existence of sin, the need of superhuman aid, the salvation

---

* It is needless perhaps to remark that attempts have been made to discover trinal analogies in nature, such as the threefold dimension of geometric figure, &c. Such attempts were made in the Neo-Platonic School of Alexandria, and in England in the last century. Most persons very properly reject them as mystical and unreal.

by mediation, the dignity of sacrifice—all these truths, though heightened and explained by revelation, yet are written in the scheme of nature, and intertwined with the tissue of the visible creation. But when we transcend these, and pass from the work to the agent, from the government of God to the mysterious nature of God Himself, we are lost in mystery; speculation is well-nigh hushed before the overpowering glory of the Eternal. We pass from the earth to the heaven, we enter the shrine of the Divine presence. We contemplate in spirit the mystery hidden of old, the mystery of the trinal existence of Him who is the source of all power, the first cause of all creation; Him who, in the depths of a past eternity, existed in the mysterious solitude of his Divine essence, when there was still universal silence of created life around His throne, and who will exist ever in the future of eternity, from everlasting to everlasting, God.

Speculation is, on such a subject, vain; yet a reverent attention to that which has been made known to us is our fitting duty. And nothing will more completely prepare us for considering the subject in a proper temper than the reflection that this great doctrine is not revealed to us in the Scripture to gratify our curiosity, but as a practical truth deeply and nearly related to our eternal interests, not in its speculative but in its practical aspects. For you should carefully note that the doctrine admits of these two distinct points of view. It may be looked at speculatively, as unfolding the nature of God; and then it becomes the battle-ground of weary controversy, and men doubt it, or misunderstand it, or add to it in the hard logical formulas which are necessary to give precision to human ideas; or it may be looked at practically, as showing us three distinct relations which God is pleased to sustain towards man, and three corresponding classes of duties which man is under obligation to perform towards God. This latter, or the practical aspect, is the view under which the subject is presented to us in the New Testament; the former, or the theoretical aspect, is that under which it has generally been regarded in the history of the Church. The Bible contains the practical doctrine, the Athanasian Creed the speculative. It is easy to perceive that the practical view is immensely the more useful; and happy should we be if we could lay aside controversy, and simply believe. But we can never hope to do so; and, therefore, it becomes important to form to ourselves definite views on the speculative controversy, for we cannot, in this age, receive the kingdom of God literally as little children. We cannot, if we would, ignore the controversies which have gathered round Christian doctrines in the course of eighteen centuries; we cannot think of those doctrines apart from the ideas which have crystallized together with them. We cannot think on all subjects of life and science with the healthy, critical, inductive spirit of the nineteenth century, for six days of the week, and lay aside our habits of thought in the church on Sundays, to receive truths with the simplicity of Jewish believers, or the reverence of mediæval mystics. We gaze on the rays of truth which come forth from the eternal source of glory in Christ and from the Pentecostal fire; but those rays come to us piercing through the distance of eighteen centuries, tinged in their passage through the mists of human thought; and we cannot hope to view them in their purity, without using rational means to deprive them of their tinge of color, and to destroy the width of their refractions.

And, therefore, I should hope that we shall not misemploy our time on the present occasion, if we restrict our attention to the speculative side of this great doctrine. It is possible to make it clear, perhaps also to make it interesting. And we shall be likely to secure both results if we sketch briefly the progress of thought in reference to this doctrine through the Christian history, noting one or two great epochs, when Christendom has been agitated by the controversies respecting it; controversies which have left their impress on succeeding ages, and live still in the hearts, if not in the creeds, of men and churches.

We must assume (for in the few minutes of the present discourse we cannot pause to prove it) that our Blessed Lord taught, and that the Apostles intended to convey the doctrine, that the Divine nature consists of three distinct classes of attributes, or (to use our human expression) three personalities;* and that each of these three distinct Persons contributes separate offices in the work of human salvation; God the Father pardoning; God the Son redeeming; God the Holy Ghost hallowing and purifying sinful men.

This doctrine can, I believe, be proved distinctly from the New Testament; and history can be added as an attestation to show that it was the primitive teaching of the

---

* *Personality*, as is well known, is the translation of the term ὑπόστασις. It is hardly necessary to remark that the writer of this Sermon does not intend—by the use of cautious modes of statement, such as the one which is here given in the text, and similar ones which occur afterwards—to favor the Sabellian theory of the Trinity, which made the distinction of the three persons to be subjective instead of objective; distinctions in the mode of God's revealing himself to man, instead of real distinctions in the Divine nature. Such a theory is precisely an instance of those very attempts to venture beyond the teaching of Scripture, against which this Sermon is designed as a protest.

Church.* Nor does it seem that for two centuries and a half any doubts were felt on the question. Christian doctrines were, indeed, during those two centuries, brought into contact with heathenism, and many of them underwent free criticism; but controversy did not invade the doctrine of the mysterious existence of the Holy Trinity. It was in Egypt that the controversy awoke, in the city of Alexandria,—that city, planted by the Greeks, which was at once a mart of commerce and a seat of learning; the meeting-point of East and West, alike in manners, in religion, in philosophy. It may seem unfit to bring before you historical allusions which might be judged ill-suited to the pulpit, but the truth is that no one can, without knowledge on this subject, understand the Athanasian Creed, which we occasionally repeat, and which, when we do understand it, you will perhaps agree with me, there is no cause to wish to have removed from our service books; and also there is another reason which it were affectation to ignore, which may justify me in touching on this subject. Any one who knows our popular literature will be aware that within recent years the thoughts of those old Alexandrian thinkers, who fought against our holy religion, and who forced upon Athanasius and the noble band of Christian defenders the very weapon of logical terms by which they sought to overthrow it, have been made familiar in some of the most talented works of fiction which our age has known. One of the most brilliant minds which England's church can boast at this moment† has consecrated

---

* Perhaps the most complete statement of the evidence in favor of the doctrine of a Trinity is to be found in Vogan's "Bampton Lectures."

† Rev. C. Kingsley, Jr. See especially his "Hypatia;" and also his "Lectures on the Philosophers of Alexandria."

his great powers of imagination to portray the character and reproduce the thoughts of that martyred woman, in whose death the heathen system of philosophy in Egypt was extinguished; and, therefore, in alluding to such subjects I may fairly presume that you are not strangers to them; in fact, I am only carrying out a practice to which our sermons ought really to conform, of conveying religious information addressed to the thoughts of common life, just as lessons on religious duties ought to be adapted to the difficulties of ordinary employments.

It was from about the 3d to the 6th century of the Christian era, in a time when the storm of war or civil commotion had almost destroyed the other great seats of education, that the Greek University at Alexandria offered a retirement for the thoughtful and the speculative. It was then that there arose a school of philosophers (vulgarly called Neo-Platonists) who combined Eastern and Western modes of thought. Inheriting from the Greek thinker, Plato, that power to mount into the world of abstractions, that power of transcendent genius which led him to outstrip his age, and almost to think, as it were, in modern ideas;* they also inherited the Pantheistic spirit,

---

* This modern aspect of Plato is usually thought to exist, not merely in the political problems which occur in his treatises of the "Republic" and "Laws," but mainly in his contrast between the fixity of the νούμενον or ίδεα, known by the reason, and the fleeting character of the φαινόμενον, known by the senses; which, translated into the language of modern philosophy, is the contrast between the immutability of Nature's laws and the mutability of Nature's phenomena. This modern aspect is, however, more apparent than real. In truth, all Greek philosophy, anterior to the Stoics, is ancient; and the forms under which it was presented by its authors are more or less obsolete. It was in them that the modern element was first developed. It is gratifying to be able to refer to one work in which the historic development of philosophical

which had outlived the decay of the old Hieratic system of Egypt, and the mystic, allegorizing tendency, which, springing up in the East, and raised into a system by Philo,* was the means of distorting plain truths of fact or of doctrine into mystical meanings which their authors intended not. This school of thinkers denied the pretensions of Christianity to be a Divine revelation, and attempted to establish philosophy as a rival to its claims. They were not the first enemies that Christianity had encountered, but they were the first educated men who had carefully examined and rejected its claims. Their history† may be told in a few words, though it embraces about three centuries, commencing approximately from the year A.D. 200. It divides itself into three epochs. In the first, the movement was a *metaphysical* speculation; in the second, a *political* organization; in the third, a *logical* system.‡ During about a century and a half, the ideas were gradually evolved from those miscellaneous sources which I have just now indicated. Plotinus is the great writer of

---

thought is really kept in view,—viz., Sir Alexander Grant's edition of Aristotle's "Ethics" (especially vol. i); and his "Essay on the Stoics," in the "Oxford Essays for 1858."

* Probably, however, Philo did not stand alone, but was merely one of a school, of which the other writers are now lost. See the essay on the Philosophy of Philo, in vol. i of Professor Jowett's work on "St. Paul's Epistles."

† The materials for their history, besides the study of their writings, and of Creuzer's Proleg. to Plotinus's Ennead., are to be found in Gibbon ("Decline and Fall," ch. 23); in the two works by Kingsley, to which reference has been already made; in Lewes's "Biographical Hist. of Philosophy;" in Maurice's "Hist. of Philosophy;" and in Donaldson's "Hist. of Gr. Lit." (ch. 53, 57.)

‡ The first movement is from about A.D. 200–360; the second, 360–363; the third, 363–550.

this first phase of the intellectual movement; and the wide effects of it may be seen even within the precincts of the Christian church in the system of Gnosticism. The second, or the political movement, is contracted to the narrow space of the reign of the Emperor Julian. It was the attempt to carry out the views of these unbelievers by political measures which must ever give an historic interest to the brief, the brilliant reign of that emperor. Educated in the opinions of these philosophers, and fortified with an intense dislike to Christianity, Julian felt that it was not enough to persecute our holy religion, but that an attempt must be made to disprove it, as well as political inducement held out for uprooting it. The sudden death of Julian in the Parthian wars put an end to this system; and it was not merely in vexation, but in the despairing conviction that his death would cut short the great object of his life, and that Christianity would henceforth triumph unimpeded, that, when wounded in the battle-field, he died, exclaiming, "Thou hast conquered, O Galilean!" Such was the fate of the *political* movement. The third phase of existence of the school to which I have alluded is found when, in the century after the death of Julian, Proclus attempted to recommend his ideas to the convictions of men by investing them with the rigor of a *logical* system.

It will be sufficient to have indicated thus much of their history.

Now among the many doctrines which this school of philosophers assailed, one of the principal was the doctrine of the Holy Trinity. They did not attempt to deny it; but they tried to show that it was only an imperfect attempt to express that which their science could detect by unassisted reason, viz., that Deity must exist in three states, —as simple existence,—as intelligent existence,—and as active creative existence.* The first was the Father; the second, the Word; the third, the Spirit. Thus, to them the Christian doctrine was no new truth; it was rather a formula of which they professed to be able to suggest a better interpretation.

How then did the Church of that age meet this view? It met it by reasserting the doctrine as it conceived the Apostles to have communicated it. It professed that instead of attempting to penetrate the depths of the Infinite Mind, it was content to rest in what was revealed; that instead of trying to know God that it might love Him, the Church sought to love God that it might know Him. Yet obliged to meet this scepticism with a definite statement, it was compelled to array the truths of revelation in the precise dogmas of technical philosophy; and so it gave expression to its thoughts in the Athanasian Creed. It is almost certain indeed that that formula was not the composition of the heroic Christian apologist whose name it bears. It probably originated in Southern France in the fifth century, but it entirely gave expression to the thoughts of orthodox Christendom.† And it is in reference to its age that its value must be tested. I may be permitted to say, that I can have no sympathy with those who would tear it from our Prayer-book, because they test it by our modern ideas, and examine it apart from its historical position. The man would be despised who, looking on one of the crude works which marked the first revival of Art, should persist in criticising those attempts by perfect modern styles, and who could not appreciate the efforts observable to throw off the stiffness which cramped the

* These are explained in a creditable manner in Lewes's "Biographical History of Philosophy."

† See Waterland's "Critical History of the Athanasian Creed."

early movement, and to gain a knowledge of correct form, of delicate modelling, of true shading, of natural coloring, because they do not accord with the modern standard of attainment. We should claim that those works must be estimated by the light of the age in which they were produced. Similarly, surely, the Athanasian Creed ought to be estimated in reference to the circumstances which created it. It was the Church compelled to arm itself in the weapons of logic used by its assailants. It is not the form in which the Church would have preferred to record its faith;* and when we still use it, we read it as an historic memorial, a protest against heresies once prevalent and analogically applicable to ourselves; the vigorous expression of belief of men who lived and strove for the faith which they loved unto the death.

With these remarks we may leave the consideration of those early controversies of the Trinity,—controversies which however I shall show in the sequel have been recently revived, and pass on to the consideration of another, which still numbers many adherents,—the rise of Socinianism at the time of the Reformation.

A thousand years separate these two great crises of intellectual speculation. When the latter of the two arose, Europe was no longer the same. The old centres of civil-

---

* This may be inferred from the fact that the creeds became more complex as the interval which separated them from the Apostles' time increased. The earliest creed, perhaps, is in St. Paul's first Epistle to Corinth (15 : 3-8). The next is the creed which was proposed and rejected in the Council of Nicæa; the next is the creed actually adopted; the next is the one adopted by the Council of Constantinople (A.D. 381), which is the one which now passes under the name of the "Nicene Creed;" the next is the Athanasian. The authority for some of these latter statements is the historian Socrates.

ization had shifted, the old forms of government had vanished; the nations had changed, the very languages had disappeared. One power alone had survived the deluge which had changed the face of Europe, viz., Christianity. It had been the ark of refuge through that deluge. It was now to have its claims tested by the busy speculation of the new world. Nor is it to be wondered at, that, in the general dissolution of the intellectual and religious system of the middle ages which we call the Reformation, spirits should arise to explore the very foundations of our faith. It was at this time accordingly that Sozini* started the theory of modern Unitarianism, by attempting to show that the Apostles had not really taught the doctrine of the Holy Trinity, but that, on the contrary, the doctrine had arisen in the early centuries, in the course of controversy with the Alexandrian philosophers. I shall dwell for a few moments on these views, because they still linger in society, and still receive acceptance from many. The arguments which are adduced in denial of the doctrine which we believe that the Apostles taught, consist partly in a critical examination of Scripture passages; but still more in an unconquerable objection to the doctrine on account of its involving a mystery. "How," urge they, "can the Divine Being exist in three persons and in one?" On the examination of Scripture texts, it is unnecessary now to say anything; but I wish to add a few words in reference to the objection that the doctrine is mysterious, because it may have frequently suggested itself to many of you.

The fact that this doctrine involves a mystery, is so far

---

* For Socinus, and the Racovian School which arose from him, see Hallam's "Hist. of Lit.," i, 552; ii, 335.

from constituting a fair ground for its rejection, that it agrees in this respect with many of the most allowed truths of human science. For the distinction is now well understood between a truth being *apprehended* and its being *comprehended*. We *apprehend* or recognize a fact when we know it to be established by evidence, but cannot explain it by referring it to its cause; we *comprehend* or understand it when we can view it in relation to its cause. A thing which is not apprehended cannot be believed, but the analogy of our knowledge shows that we believe many things which we cannot explain or resolve into a law. We know the law of attraction which regulates the motions of the visible universe; but no one can yet explain the nature of the attractive power which acts according to this law; or, to add an example from the world of organized nature, we may see the same truth in the animal or vegetable kingdoms. We know not in what consist the common phenomena of sleep or of life; and we are equally ignorant of the final causes which have led the Creator to lavish his gifts in creating thousands of species of the lower orders of animals with few properties of enjoyment or of use; or to scatter in the unseen parts of the petals of flowers the profusion of beautiful colors. In truth, the peculiarity of modern inductive science is that it professes to explain nothing. It rests content with generalizing phenomena into their most comprehensive statement, and there it pauses; it in no case connects them with an ultimate cause. And if truths are thus received undoubtingly in science when yet they cannot be explained, why must an antecedent determination to disbelieve mystery in religion be allowed to outweigh any amount of positive evidence which can be adduced to substantiate those mysteries.*

* It is fair to state that the antecedent objections which are urged

We have now noticed the two great attacks on the doctrine of the Holy Trinity which have been marked in history; the first, that which accepted the Trinity, but explained it away; the other, that which denied the doctrine on the ground of its mystery. Yet the subject would hardly be complete if I were not to notice with a brief allusion the fact that an attempt has arisen in Christian writers during the present century, alike in Germany and in England, to revive speculations, similar to those of the old philosophers of Alexandria, in defence of this great doctrine. One honored layman, whose influence was great equally in letters and theology during the first thirty years of this century, poet, critic, philosopher, theologian alike,* has been the parent mind of a school of earnest and deep thinkers, of whom some are gathered to their home above, some still live to serve the Church on earth. It ill becomes

against the doctrine are of two distinct kinds: (1st), arising from the unwillingness to believe a thing incomprehensible, which is the one refuted in the text; (2d), arising from the impossibility of accepting a truth contradictory to reason, in believing three persons to be one, and one to be three, at one and the same time. This latter objection is of course reasonable in itself, but incorrect in its application; inasmuch as this is not the Scripture account of the doctrine of the Trinity, but the clumsy and self-contradictory statement of unintelligent advocates of it.

* S. T. Coleridge. A little book exists on the effects of Coleridge on theology, entitled "Modern Anglican Theology, by the Rev. J. H. Rigg." The friends of the writers whom he has criticised will naturally consider the book very unfair. They look at the works through the writers of them, Mr. Rigg looks at the writers through their works; hence he has certainly, in most cases, especially in that of Mr. Jowett, presented a caricature of those whose works he discusses; and has not unfrequently imputed to them as positive teachings ideas which are only to be found in their writings as incipient tendencies; but in spite of these and other defects the book is instructive.

so young a student as myself to criticise those views; it must suffice to have named them. I may venture, however, with all humility, to remark that they do not appear to me to convey any help towards elucidating this great doctrine; nevertheless, if others find that they afford them support, I would be the last to tear from them the reed on which they support themselves, frail though I fear it to be.

Having now completed the brief history of this great doctrine, having seen what we are not to believe, let us turn in conclusion to see what we ought to believe of the nature of God, and what the lessons are which we should carry away from the consideration of it.

We have asserted that we are to believe that the Divine nature exists under three entirely distinct classes of relations, which through poverty of language we call existence in three Persons. We must be careful, however, when we assert this, not to reduce the Divine nature to similarity with the human, not to commit, in fact, almost the very error into which men of old fell in supposing that the God whom the heaven of heavens cannot contain is like to birds and beasts and creeping things. The Divine Being is three Persons; but by this we only mean that the personal element in man is the analogy under which God has been pleased to convey to us ideas of His own nature and of the relations which He sustains to us. Revelation, when teaching truths of the world unseen, must of necessity be compelled to present them by comparison with things that are known. It must, therefore, select its illustrations either from the world of matter which is known to us through the senses, or the world of mind and feeling known to us through consciousness. And thus to a sensuous people like the ancient Jews, God was represented as having arms and hands, or as being swayed by human passion, by anger, hate, repentance; and to Christians, God is described, in that religion which was to commend itself to the more civilized nations of Europe, as having the higher qualities of mind, and as invested with the ineradicable and mysterious attribute of personality.*

Yet, though the conception is far nobler than the old Jewish view, we must not allow ourselves to suppose that it is more literally true. Just as we do not attribute to God a body of human passions, but merely mean that He acts to us as though He possessed them; so when we attribute to Him thought or personality, we must not narrow down the idea of his omniscient intuition by supposing it contracted within the limits of inference which govern man's finite intelligence, or gifted with that limited independence which appertains to human personality. The discoveries of science ought to teach us that we really can scarcely form any positive idea of God's nature.† If we track the infinity of creation, we see that each increased power of our instruments reveals to us illimitable profusion in creation; the telescope revealing the troop of worlds stretching to an infinity of greatness, and the microscope a world of more and more minute life, stretching to an infinity of minuteness; or when we turn from the infinite in *space* to the infinite in *time*, if we look backward we see written in the rocks of the world the signs of creative life

---

* The view here advocated is an extension to the doctrine of the Divine nature, of the interpretation of the subject of analogy, which was applied by Archbishop King to the subject of predestination. See Whately's edition of King's Sermon; Copleston's "Discourses on Predestination."

† The impossibility that a mind, constituted as the human mind is, should employ itself successfully in speculations on the subject of infinity, has been developed from another (the psychological) point of view in Mr. Mansel's "Bampton Lectures."

stretching through ages anterior to human history; or if we look forward, we can detect by delicate mathematical calculation an amazing scheme of Providence providing for the conservation of harmony in the attractions of the heavenly bodies in cycles of incalculable time in the distant future. And when, having pondered all these things, we think of the Being that has arranged them by His providence and conserves them by His power, what notion can we really form of His nature? What notion of the wonderful originality evinced in the conception of creation, what of the profusion shown in the execution of it, what of the power in its conservation? His nature is not merely infinite, it is unlike anything human, and we must turn away with the feeling that when we compare that infinite Being with man, and confine our ideas of His illimitable vastness and His inscrutable existence by the notion of the narrow personality which is delegated to us finite creatures who live but for a day on this small spot of earth, lost amid the millions of worlds which glitter in creation, we may be sure that the Divine nature as really transcends the earthly description of it, as the universe exceeds this world; and though we may thankfully accept the description of God as having three personalities as the noblest to which we can attain as men, and as enough for our present wants in this world, yet let us never doubt that really the Divine nature is vastly nobler; and let us bow with adoring thankfulness in meditating on the idea which we are permitted to attain, imperfect though it be, of that mysterious essence.

Yet though the idea of God in three persons may be held to be thus speculatively imperfect, let us never forget that it is practically all-sufficient for us. For it teaches us the great truth that He acts to us as though He did literally sustain the characters of three wholly distinct persons, and that He demands from us the duties which would belong to us if He were so.

If we are thus to believe of God, what is the lesson which this great doctrine that God exists and acts to us as Father, Son, and Holy Ghost, ought to convey to us? It is mainly the wondrous thought that this glorious Being is willing to stoop to be our friend, that He whose happiness is complete in its own infinity, is moved by His own pure eternal love to win us to Himself. Restless (to speak after the manner of men) to secure our happiness, all these blessed persons of the glorious Godhead are engaged to secure it. It is God the Father whom we have grieved by our sins; and yet he loves us as a father still; and to rescue us from our misery, He has designed the great scheme of salvation, and sent God the Son to dwell on this earth as a man, as a man of sorrows and of poverty, to remove by His atoning death the impediments which, secret perhaps to us, stand in the way of our salvation, and to exhibit the pattern of a faultless human being, that we may follow his steps; and lastly, after God the Son had withdrawn from the earth, God the Spirit, the ever blessed Comforter, has descended to dwell constantly in the hearts of all men that invite His presence, cheering their guilty spirits, stirring them up to the love of holiness, hallowing them for a meetness for the inheritance of heaven. Behold what manner of love God has shown to us! Behold the Triune God engaged in the salvation of each one of ourselves!

And can you delay to yield to Him your hearts, your wills, your affections? If you have sinned, or are tempted to sin, either in deed, or word, or thought, remember that it is not merely sin against a law, but that you are verily grieving a loving father, even the Father, God; if you are living a careless, half-religious life, remember that you are perpetrating the ingratitude of making the sufferings of the Eternal Son void as regards your souls; if you are neglecting prayer, neglecting earnest supplications to heaven for holiness, you are declining to avail yourself of that unspeakable gift of the Spirit's help which is for all that ask.

Forget, if you like, all those hard historical and logical discussions with which I have perhaps misemployed your precious time in this sermon; forget, if you please, the nobler views of God's personality to which I have striven to raise you. Think of Him, if you choose, only as three persons. But forget not that His eye is now upon each one of you, that he seeks to have each one's heart. And if, in the portion of leisure which is now afforded us,* any of us are about to go forth in quest of health or instruction to foreign lands, let us never forget that when we have passed the Straits which insulate our native land, and are emancipated from the restraints of English society and the sanctity of English sabbaths, yet God's eye is over us and His presence nigh to us. Let us never forget that in whatever scene we may find ourselves, whether lost amid the thousands of a crowded city, or halting beneath the humble roof of the mountain peasant, still none of our ways are unobserved on high; and be it our perpetual consolation that there is instant access for us to God's throne by prayer; nay, that if there be in us any good desire, He sees it ere we shape it into words, and from His invisible throne, swifter than the speed of thought, there descends the answer of love. Let each of us strive to use the leisure

---

* This Sermon was preached immediately before the commencement of the Long Vacation.

time on which we are now entering not only as a means of securing a higher mental cultivation, but also for gaining a deeper communion with the God of glory. For God the Father loves us, God the Son has redeemed us, and the Holy Spirit will, if we will ask Him, turn us from sin, and doubt, and half-heartedness, to the love of Himself, and will fit us for that heaven where, no longer trammelled by sin and darkened by ignorance, we shall enjoy the beatific vision, and find our everlasting happiness in communing with the Divine Being face to face.

SERMON VI.

THE ATONEMENT.

(PREACHED BEFORE THE UNIVERSITY, MAY 15, 1859.)

MARK 9:2.

"*And after six days Jesus taketh with him Peter, and James, and John, and leadeth them up into an high mountain apart by themselves: and he was transfigured before them.*"

THE pilgrim traveller who wanders through the land once hallowed by the bodily presence of our blessed Saviour, never fails to have his attention attracted by the sight of one hill which stands conspicuous alike by the beauty of natural features and the interest of traditional associations. The hill is Mount Tabor.* From whatever position the traveller may approach it, as he reaches the escarpment which overlooks from all sides the wide plain of the river Kishon, in which it stands, the mountain comes into view rising in queenlike majesty from the surrounding plain. Standing alone like an island in a sea, with its ragged slopes rounded into a conical outline, it presents the appearance of cheerful fertility; aged olives, with their trunks

---

* Stanley's "Palestine," pp. 343–392.

gnarled by time, besides other trees, are dotted over its surface, while the outline of rude fortification, almost coeval with our Saviour's life, is discernible, forming the coronet of its summit. We cannot wonder that ancient tradition should have selected this spot as eminently "the mountain apart," to which some of the Evangelists allude in their narrative of the scene of the Transfiguration; we cannot wonder that those whom the superstition of a pilgrimage, or the excitement of a liberal curiosity, has at various times attracted to the spot, have looked with uncommon emotion on a mountain which to its natural beauty added the interest of supposed connection with one of the most marvellous, the most poetical passages of our Saviour's earthly career.

It is cruel to dash away such a belief; yet the rigor of geographical criticism compels us to doubt whether that spot can be the real scene of this event. For the comparison of the narrative of St. Mark with that of the other Evangelists shows clearly that the words, "He leadeth them into a high mountain apart,"* do not refer to the mountain, as if the mountain were described as standing "apart," but merely meant, "He leadeth them apart into a mountain," *i. e.*, "apart" from the hurry of men, "apart" from the other disciples to whom it was not vouchsafed to gaze on the mystic vision. Accordingly, if we look with care into the history, we find that our Lord was, at the time of the event of the Transfiguration, far removed from Mount Tabor, in the hilly district either of Ituræa, or of the northeast of Galilee.

Mount Tabor will still continue to be a spot of interest alike to the man of taste and to the student of history;

---

* Mark 9 : 2.

and the traveller will still turn aside to climb to its summit, and gaze upon the panorama outspread to his view; but as he looks around him he must content himself with the remembrance of the stirring events of which that scene has been the undoubted witness, without associating the wild beauty of the mountain summit with the sacred history of our Blessed Lord. That hill, if it had voice, could testify to many an exciting scene which has taken place in the plain at its base. It heard the shout of victory which Deborah raised over the discomfited host of Sisera. Not far from it occurred the midnight panic which the hero Gideon struck into the camp of the Midianitish hordes. Hard by stood the hills where Saul fell by the hand of his armor-bearer; and from age to age, down to the memory of living men, it has witnessed the surging waves of successive invasions roll around its base; but the marvel of the Transfiguration it cannot have witnessed. The scene of that event must be sought elsewhere, far off in the solitudes to the northeast of Galilee. There, in some unknown spot, amid the spurs of the snowy range, from whose roots bubble forth the sources of the river Jordan, was enacted that marvellous event, when Jesus was transfigured before His startled disciples. We may regret the impossibility of knowing the exact spot; we may be sorry to be obliged to think of the event without connecting it with the place; but we really lose nothing by the circumstance; for the lessons taught us by the Bible are spiritual, not temporal; moral, not geographical;* the conditions of place or of

---

* Some observations will be found in Mr. Jowett's work on "St. Paul," (vol. i, pp. 29–31, 1st ed.), on the comparative unimportance of geographical and historical knowledge as means of gaining a knowledge of the minds of the Scripture writers. It aids us in reproducing the external scene, but does not penetrate into the internal spirit.

date matter little; the truth taught is eternal; and we, who in the peaceful quiet of this church are now turning our thoughts to that striking event, may realize its eternal truth, and gather its eternal lessons more really than if we could determine the site of its occurrence, or kindle our sympathies by a pilgrimage to the spot.

Yet if it were our object to find interest in the mere narrative, and poetry in the mere scene, the real site, amidst some craggy fastness of northern Palestine, would supply it hardly less vividly than the beauteous form and picturesque situation of the solitary Tabor. For we should realize to ourselves the fact that our Blessed Saviour, after having exercised His ministry of mercy for more than two years,* was now a fugitive, wandering near the frontiers of the land of Tyre and Sidon, in order to escape the observation of the spies sent down by the authorities of Jerusalem to track His footsteps; and that, when thus circumstanced, He one day withdrew into a mountain, accompanied by his three favorite disciples, Peter, James, and John, to pray. There, while the disciples were left sleeping, while He himself was drawing nigh in prayer to His Heavenly Father, and His Heavenly Father drawing nigh to Him, the powers of the eternal world, over which He had reigned ere He came to this earth, broke in upon Him; the glory which He had with the Father before the world was, overshadowed Him; His face shone as the sun; and His raiment became white as the light. The veil which hides the unseen was lifted, and strange visitors from that world stood beside Him. The radiance of that glory broke in upon the still

---

* The best harmony of the Gospels which the writer of this Sermon has ever met with is that by the traveller, Dr. Robinson. It is published in an English form very cheaply by the Religious Tract Society, and is accompanied with very judicious notes.

slumbers of the beloved disciples, and as they woke, a bright cloud was overshadowing them, and they feared as they entered into the cloud. And a mysterious voice was heard, which said, "This is my beloved Son, in whom I am well pleased. And when they had looked round about, they saw no man any more save Jesus only with themselves."*

Such was the scene. We need not surely pause to prove that it was not a mere dream of the three Apostles. This idea is forbidden by the fact that, in imparting the narrative of the transaction to those evangelists who have handed down the history of it to us, they have exactly marked the line which separates that part of the vision which was taking place, when in the half-unconsciousness of persons awaking from sleep, they hardly knew what they said, from that portion which occurred after they had become cognizant of events around them.† Above all, such an idea is forbidden by the fact, which will be more fully explained presently, that from that moment our Saviour's teaching completely changed; He began henceforth to divulge the fact of His own sufferings.‡ An event must have been real to Him which forms an epoch in His teaching and His life. Resting accordingly with confidence on its reality, believing that it was not a mere sleeping dream of the Apostles, but a great though mysterious fact, that at that moment heaven and earth were brought nigh, and Jesus transfigured in celestial brightness, and visited by spirits from the world unseen, and that the Apostles were permitted to catch a glimpse of the last rays of that departing glory, let us ask

---

* Matt. 17: 1–13; Mark 9: 2–10; Luke 9: 28–36.
† Mark 9: 6; Luke 9: 32.
‡ Matt. 17: 22, 23; 20: 17; Mark 9: 12, 13; Luke 9: 44.

ourselves what was the purpose and intention of that wondrous revelation of things from within the veil? What was its meaning? what relation had it to the Apostles? what relation had it (we ask it without irreverence) to the Saviour?

It has been common to suppose that it had no meaning to the Saviour himself, but that its sole object and purpose was to instruct the disciples; accordingly it has been thought to be a parable, as it were, *acted*, in order to figure, by the meeting of Moses and Elias with Christ, the union of the old dispensations of the Law and the Prophets, of which they were distinguished representatives, with the new, which Christ came to proclaim. A moment's consideration, however, will at once lead us to perceive that such an interpretation is paltry, and utterly unworthy of the magnificence of the scene. God does not thus act either in nature or in religion; he does not inaugurate mighty agencies to usher in insignificant results. A deeper meaning must therefore be sought; and it may easily be found if we look a little more closely into the evangelists' narrative.

St. Matthew and St. Mark alike inform us* that, when the vision had ceased, our Lord charged His disciples that they should tell it to no man till the Son of Man should be risen from the dead; and that it gave rise to a declaration by our Lord of the sufferings which He should shortly undergo. And St. Luke adds a fact which explains why the conversation had turned to this subject, when he says that Moses and Elias, as they appeared in glory, spake of Jesus' decease which He should accomplish at Jerusalem.† So it was, it appears, the subject of our Blessed Saviour's

---

* Matt. 17:9; Mark 9:9. † Luke 9:31.

coming sufferings which was occupying the thoughts of those visitants from the other world; it was this which brought them back to the earth which they had so long left; and it was some of the last of those notes of sadness which the disciples heard as they awoke from their sleep, which led them to enter into discourse with our Saviour on the subject of His sufferings, and which caused Him to charge them, as though they had been permitted to learn a secret of the coming time which was not to be divulged to the uninitiated, to tell no man until the Son of Man should be risen from the dead.

If we would fully understand the bearing of this conversation about our Lord's sufferings, we must also take into account that strange fact to which allusion has been previously made, that from the time of the event of the Transfiguration, our Lord's teaching underwent a change, in that henceforth He announced that He was to suffer and to die. Heretofore He had not announced this. He had taught that He was the Messiah; henceforth He taught that He was the Messiah who was to suffer. Heretofore He had taught that He had come to fulfil the predictions of the ancient Hebrew prophets, to found that kingdom of power on which the hopes of generation after generation of the Jewish people had been set, and after the glory of which their longing eyes were straining, as they went to the grave unblessed by the expected day of liberty which was to free them from the nations which oppressed them. Henceforth He taught them that He was to set up that kingdom by means of Himself suffering, that He was to fulfil in Himself the affecting descriptions of the prophets: "He was cut off from the land of the living; for the transgression of my people He was stricken."*

* Isai. 53 : 8.

The reason why our Lord had hitherto kept back this doctrine is not hard of discovery. It was not as one of the rationalistic critics of Germany* has suggested, that He found the Jewish people so brutal, so unapproachable by moral teaching, that He began to adopt the plan of trying to reach them by an appeal to sorrow, and to court death Himself to attest the honesty of His own teaching. The real cause was this: that our Lord first taught morality before He taught faith. He first taught His hearers to listen to Moses before He expected that they would listen to Him; He first taught them to arouse themselves from sin by repentance, before He communicated the intelligence of His own sufferings, which were to be the means of redeeming them from the misery of sin. Therefore, after He had condescended to devote more than two years of His ministry to proclaim Himself the Messiah, the Messiah who wished to gather round him reformed men and honest hearts, He devoted the remaining portion of hardly a year to proclaim Himself the Messiah who was to suffer,—the Messiah who, by His death, was to make an end of sin, and to bring in an everlasting righteousness.

From this point of view we can in some humble measure see what was the purpose and meaning of the mysterious event of the Transfiguration. It was not merely, we venture to think, meant as a lesson to the disciples, a lesson to teach them the importance of Christ's sufferings, to inform them that those sufferings occupied the attention of departed prophets and of inhabitants of the world unseen, nor to convince them of the glory of His character; but it had a real use and meaning also in reference to our Blessed Lord. It taught the three disciples these truths

---

* De Wette, "De Mort. Chr. Exp." ii, sect. 23.

indeed, and doubtless the lesson was never forgotten. In reference to St. James, we know it not; but in the case of St. Peter, we have his own attestation, shortly before he was about to put off this tabernacle, that he had been an eye-witness of Christ's majesty, when he had seen His glory on the holy mount.* And in the case of St. John we may well conjecture, that when, a generation later, he wrote his Gospel in extreme old age, and declared—"The Word was made flesh and dwelt among us, and we beheld His glory, the glory as of the only begotten of the Father,"† —the thoughts of the Apostle were travelling back, across the memories of sixty years, to the scene which he had once witnessed amid the bleak cliffs of the Syrian mountains, when the face of Jesus had shone like the sun, and His raiment glistened as the light.

While, however, the event thus instructed the disciples, we cannot doubt, from what has been before said, that it had a real use and meaning also to our blessed Lord. We are wrong in levelling the mystery of those few events in our Lord's life, when He was overshadowed by the powers of the world unseen, to the standard of our pigmy explanations. In that outburst of temptation, for example, which He endured in the wilderness at the commencement of His ministry, there was doubtless, beside and above all that is common to man and an example to man, a real vicarious and mysterious endurance of temptation by our Lord as part of the system of mediation which He had undertaken. And in that mysterious agony of dread and terror which befell the Saviour in the olive garden of Gethsemane, it was not under the pressure of ordinary mortal horror that He was bowed down; we may well

---

* 2 Pet. 1 : 14–18.   † John 1 : 14.

believe that there was then going forward between the soul of our Divine Redeemer and His Heavenly Father some secret transaction connected with human redemption, that He was verily drinking our cup of sorrow, and sweating drops of blood in the vicarious endurance of our load of sin, that it was the weight of the sins of the world under which He was staggering which made Him breathe out, in the exhaustion of His agony, "If it be possible, let this cup pass from me."*

In like manner with these two events of the Temptation and the Agony in Gethsemane, we may well believe that the similarly mysterious event of the Transfiguration had a meaning to Christ over and above that which it had for the Apostles. We may not hope to understand that meaning; yet, if we may conjecture, we can conceive that that moment was the baptism into His life of suffering. We can imagine that the spirits were sent forth as heralds to tell Him of the interest with which his work of atonement was regarded in the world invisible. We can conjecture that the reason why angels were not chosen to convey that message to Him, but rather the spirits of departed men, such as Moses and Elias, was, because they were spirits who had tasted of human sin, and whose welfare depended, together with that of the universe of created men, upon the work of suffering which Jesus was to commence. We can imagine that the vision of Divine glory which was vouchsafed to Him, the momentary taste of the heavenly state, was to strengthen Him for that work, and to cheer Him

---

* This view of the meaning of our Lord's Temptation and Agony is worked out with equal talent and pathos in the two last Sermons of a scholarlike and suggestive volume by the present Dean (G. H. S. Johnson) of Wells. Compare also Newman's "Sermons to Mixed Congregations" (Sermon on the Agony).

with the prospect of the glory which he had left, and to which He was to return. And we may possibly conjecture, finally, that the voice from the excellent glory, "This is my beloved Son in whom I am well pleased," was again uttered in order to anoint Him for His mission of suffering, as once before it had been uttered at the baptism in the Jordan to anoint Him for His mission of teaching.

More than this we must not conjecture; yet thus far, I trust, the narrative of the evangelists, as we have presented it, warrants us in venturing. More than this we need not know; for the subject has taught us a lesson large enough already, if it has fixed our thoughts on the sufferings of Christ, and made us realize the fact, that it is not the *life* but rather the *death* of Christ which is important as the means of our salvation; not His life as an example, but His death as an atoning sacrifice.

It is so much the more important that we should feel this, because in recent years the belief has been fast spreading in our church that the death of our blessed Lord was not for the purpose of reconciling God to man by removing the obstacles, known or unknown, which stood in the way of human salvation, but only for the purpose of reconciling man to God, by proclaiming to mankind God's love, and by exemplifying the majesty of suffering and the dignity of self-sacrifice.* The assertion of this view has been a

---

* The view that the Atonement only reconciled man to God appears to be that of Mr. Jowett ("Essay on Atonement," in his work on St. Paul, vol. ii, 1st edit.); that it was a great example of self-sacrifice is the view of Mr. Maurice ("Essay on Atonement"), and of the late lamented Mr. Robertson (i, 9; iii, 7). Abelard's view was also similar. The encomium bestowed a few lines lower is meant to include the whole body of educated theologians in the present century who have,

natural reaction against the rash statements in reference to this great doctrine which have not unfrequently been put forth: statements which venture beyond the cautious teaching of the inspired Apostles on the subject, and are constructed in ignorance of the development which the doctrine has received in the progress of the history of the Church. Such statements, in which men dare to measure God's infinite nature, and assert that the reasons can be fully understood why God was unable to forgive man without an atonement, frequently run counter to those ineradicable instincts of justice and truth which the God of nature has planted in the moral sense of man, as the unassailable foundation of moral truth against which shallow, humanly-invented systems of theology dash themselves in vain. Accordingly it has been from the high and noble motive of opposing these rash theories, that the view to which I allude has been advocated. It has met with the approval, since the commencement of the present century, as will be known to many of you, of men who have been, or who are, the honor of our church, the ornaments of our universities, the glory of our literature,—men whose high moral and intellectual worth is beyond praise, and from whom I could not have felt it right to dissent without this tribute of respect in honor of them. If, however, we have reason to believe that an alarm at the rash speculations of men has led them into an abandonment of the hallowed doctrine which was inculcated by our blessed Lord and his Apostles, which has formed the piety of ancient saints, and cheered the hearts of guilt-stricken men, we may venture, without any charge of arrogance, to dissent from their criticisms,

more or less, formed their views by the study of Coleridge or of the modern literature of Germany.

and to cling tenaciously to the faith once delivered to the saints.

It is impossible, in the few remaining remarks of this sermon, to discuss this subject fully, yet I wish to point out the successive notions which have been held in different periods on the subject of our Lord's sufferings and atonement, in order that you may exactly understand the position which the view that I am opposing holds in the history of the doctrine, and may be better able to perceive the point where human speculation has failed, at which reason must expire in faith, and theory in adoration.

To the minds of the Apostles the subject was beset with no difficulties. Trained in the idea that God was in some inexplicable way approachable by sacrifice, they beheld in the death of our blessed Lord the realization of their religious aspirations, the fulfilment of the Jewish sacrificial types, the true offering for human guilt.* They believed, they theorized not; but they prayed, they put their trust in Christ's merits, and in the depth of their own religious consciousness they realized the joyful experience that their characters were changed through the power of that death; and in the gladness of pardoned hearts, and the energy of renewed wills, they went forth to proclaim to others the blessedness of which they themselves had been made partakers.

But time passed on. The Apostles one by one were gathered to their reward, and the voice of inspiration was finally hushed in the tomb of the last Apostle. There was no longer any open vision, and, in the centuries of thought and of criticism which succeeded, men began to question the cause of the atoning death of Christ. They rested no

---

* See especially the Epistle to the Hebrews (ch. 5–10).

longer in the mysterious fact, the reality of which was re-echoed in their heart of hearts, but they sought a reason why an atonement had been necessary.

And in those many centuries which intervened from the third to the eleventh age, no better answer could be found than that it was for the purpose of effecting the ransom of mankind from the Evil One that Christ had yielded up His life.* The darkness of the human intellect in those ages had clouded even Christian truth, and the very patriarch whose religious zeal led him to send his missionary to Christianize our Anglo-Saxon forefathers, and whose piety is attested by the solemn Litany which we still repeat in our services, which was first chanted by him amid the depth of the Roman pestilence, could find no nobler solution of the reason of our Blessed Lord's mysterious death than the degrading idea that it was a price paid to the Devil for the redemption of man; that as men had become the subjects of the Evil Spirit in conquest, Christ's death was necessary to purchase, as by a ransom, the emancipation of the human family, and their restoration to their lawful king.†

The darkness of that night of thought passed, and the glad twilight of the morning of mental illumination began to lighten up the scene. One there was whose name is

---

* This strange fact has been brought to light by Dr. Thomson, in his able "Bampton Lectures" (for 1853) on the Atonement. See Lect. vi, p. 155, and the notes.

† Gregory, in Evang. ii, Hom. 25 (quoted by Dr. Thomson). This idea of "ransom" was hardly, indeed, more opposed to the moral sense than are many more recent theories; yet it was a degrading notion to assign to the Evil Spirit a duality of power with the Divine Being,—an idea borrowed through Manicheism from the duality of Eastern religions.

usually identified with political events in the history of one of our own Norman kings, but who is honored by the student for the nobler labor of having consecrated his gifted intellect to the explanation of the doctrine of the atonement. "I do not seek,"—these were his words,—"I do not seek, O Lord, to penetrate Thy depths; I by no means think my intellect equal to them; but I long to understand in some degree Thy truth, which my heart believes and loves. For I do not seek to understand that I may believe, but I believe that I may understand."* England has the honor of possessing his remains, and amid the long series of monuments which mark the last resting-place of many illustrious men in the metropolitan cathedral of Canterbury, there exists none dedicated to a nobler memory than the stones which indicate the spot where formerly stood the shrine of Archbishop Anselm.

Anselm swept away the idea that the atonement was a ransom paid to Satan, and substituted the idea that it was a debt required by the broken law of God. The doctrine was thus no longer illustrated by an analogy borrowed from conquest, but by one borrowed from the forms of justice. Yet even Anselm laid more stress upon the holiness of Christ's life as the fulfilment of the requirements of a broken law, than upon the atoning character of His death as the very essence of that mighty mystery. It was a thinker in Central Italy, in the early part of the 13th century,—a mind which was one of Nature's prodigies, which still shines across the distance of so many centuries like a star of the first magnitude,—that threw this additional beam of light on the doctrine. I allude to Thomas Aquinas.† Nor has

---

* Anselm, "Proslog.," i, p. 43. His work is injured by mystical ideas of numbers, &c.

† "Aquinas Summa," P. iii. quest. 48.

any further conception been subsequently added to the doctrine in the progress of time, save that, as the opinions of men on the theory of human punishment have improved, growing gradually to perfection through the successive stages of retaliation, compensation, retribution, correction, till the corrective idea of it has been finally substituted in place of the vindictive, the doctrine of the atonement has been conceived less under the idea of a vengeance which righteous justice demanded, and more under that of a punishment administered for some undiscoverable reason as a mighty spectacle of correction for sin, alike before angels and before men.*

Thus the doctrine has been traced down the path of history. We have seen that the fact only was presented by the Apostles, and that successive theories have been

---

* Compare Erskine's "Internal Evidences," and Dean Conybeare's "Theol. Lectures" (Lect. iii. p. 358, &c.) See the notes to the seventh lecture of Dr. Thomson's work, before quoted, from which several references have been taken.

The various views on the Atonement since the inspired teaching of the Apostles may be classed as follows: (1st). The allegorizing doctrine of the Alexandrian Fathers, reproduced by John Scotus Erigena, in the 9th century; (2d), the Patristic view of ransom from the time of Irenæus to Anselm; (3d), that of the Schoolmen, Anselm and Aquinas; (4th), the Protestant theory of Calvin and Grotius; (5th), the Socinian; (6th), those held by the modern Germans since the revival of speculative philosophy. Materials for this history are partially supplied in Baur's work, "Lehre von der Versöhnung," above-named. He considers the tendency of theory on the Atonement to have been toward speculations into its objective nature until the Reformation; into its personal or subjective relation to the human soul from the Reformation to Kant's time; subsequently to whose teaching, he regards the doctrine as again travelling in an objective direction. A brilliant sketch of this history is given by Mr. Jowett in the second edition of his work on St. Paul (ii, 568-585).

attempts to explain it by reason, or to draw out the meaning of Scriptural statements. Yet, though we thankfully accept all the help which the ideas of sacrifice, or exchange, or ransom, or debt afford us, which of us is there that does not feel that there is some still deeper mystery unexplained, and that those theories are but feeble attempts to grasp that which transcends the powers of human cognition,— feeble attempts to present under the miniature of human analogies the magnificence of infinite mysteries?

We need only look for a moment at those astonishing contributions to the evidences of the greatness and goodness of God which knowledge is daily collecting from the works of Nature; we need only read that sketch of the "Cosmos" of the physical creation, which is the last legacy bequeathed to the world by the ripe and honored age of the patriarch of science whose remains have this week been attended to their grave by the regrets of educated Europe, to be convinced that God's thoughts are not like man's thoughts, that "the measure thereof is longer than the earth and broader than the sea." When, for example, science traces the infinity of vastness in the worlds and systems of worlds which roll around God's throne in twinkling myriads, each knowing its appointed course; or descends into the proofs of His power in the infinity of minuteness revealed in the world of microscopic life; or stretches back into the unknown depths of the past, and deciphers the mighty movements and incomprehensible purposes of the God of creation from the memorials of fossil life inscribed on the rocks of the globe; or unfolds the magnificence of His power and the vastness of His resources in the ever restless flow of causation which marks the present; which of us can fail to meditate on those mysteries with reverence? which of us can fail to feel that the unfathomable depths of that in-

finite mind are not to be comprehended by the finite powers of this being who lives for a day on one insignificant planet, an atom amid the mightier worlds that rule in creation; a being whose knowledge, where it is not simply subjective, is at least wholly relative, whether it be gathered from the range of earthly experience or from a divine revelation, which, in order to be understood, is necessarily cramped by terms and thoughts borrowed from earthly analogies?*
" Canst thou by searching find out God? canst thou find out the Almighty unto perfection? It is as high as heaven; what canst thou do? deeper than hell; what canst thou know?"

Which of us can fail to feel that such unworthy notions as barter and purchase, and substitution and insolvency, borrowed from the small range of mundane occupations, however useful they may be as aids for illustration, are not to be regarded as the full measure and entire explanation of that magnificent system of atonement which is revealed in the death of Christ; the meaning of which analogy would lead us to believe as much to exceed all our human conceptions of it as the universe exceeds this little globe?†

I am far, indeed, from wishing to tear from any mind the illustrations which it may find useful in explaining to itself the mystery of the atonement. As we are unable to form a notion to ourselves of a Divine Being, save by conceiving Him to have a body, or passions, or thoughts like

---

* Knowledge is dependent either on innate *forms* of thought, or on *matter* suggested through experience. In the former case, it is subjective in its character. A revelation must be accommodated to the powers of the being that is to apprehend it; else it would be unintelligible.

† See this line of thought pursued in Sermon V, p. 164.

those which belong to man ;* so perhaps we may be unable to imagine to ourselves atonement without the aid of earthly illustrations; yet we ought to remember that they are, at most, probably only types, miniatures, distant analogies of a reality which passes man's comprehension. We may rest in these now; hereafter, perhaps, in the heavenly world, the mystery of wondrous love shall be unfolded to us, not in blind glimpses and indistinct types, but eye to eye, by the light of an undimmed intuition, when we shall know as we are known.

Must we, however, in the consciousness of the comparative imperfection of all our explanations of this great atonement, throw away totally the mystery of it, and regard it, as we have said that many now consider it, to be only a great example, preaching the evil of sin and the dignity of sorrow? Are we to think that it only reconciled man to God, and not also God to man? Are we to suppose that its sole object was in reference to man, and that it contained no deeper mystery unrevealed to us, in reference to God and to the world invisible? Are we to believe that no obstacles stood in the way of our salvation except those understood by us? By no means. On the contrary; though I have endeavored to put you on your guard against supposing that the trifling explanations which are often given of this mighty mystery really explain it; though I have hinted that any theory of the atonement is perhaps impossible, any explanation of that majestic mystery almost irreverent; though I have preferred to advise you to rely on Christ's death as a real but incomprehensible mode of

---

* A growth might be shown in the conception under which the idea of God is presented. In the Law, God is regarded as having the *body* of man; in the Prophets, his *passions;* in the Gospels, his *mind.*

removing the obstacles, which, known or unknown to us,* stood in the way of our salvation; yet I wish to caution you very solemnly against accepting any suggestions for explaining away that atonement by making it merely the means of reconciling man to God, and not also God to man. The reasons why we cannot admit such suggestions shall be stated soon. Previously, however, it is necessary to allude more distinctly, even at the risk of repetition, to this modern form of the controversy.

It is the opinion of many conscientious men that God was never angry with man, but that man had, in consequence of human sin, begun to doubt of God's love; that Christ, therefore, came forth as a messenger, not to reconcile God to man, but man to God, by embodying a visible proof of God's love to us; and if the contrary view seem to be implied in the Holy Scripture, they explain it away, either by supposing that Providence has permitted phenomena to be therein described from the popular point of view, just as the alteration in the position of the earth is, in popular modes of speaking, attributed to a movement in the solar orb; or else that the scripture writers were presenting the idea under the Jewish conceptions of sacrifice which had trammelled their early education.†

What is the answer to such a view? It is evident that

---

* The reasons for assuming that obstacles "unknown" to us may have impeded human salvation are: (1st) because the finite mind of man is not a perfect gauge of the infinite God; (2d) we have seen that no hypothesis, in reference to these mysterious obstacles, offers a perfectly satisfactory explanation of them: and (3d) the idea of *guilt* appears to point to an impediment in Deity external to man, and not merely to the subjective or internal obstacle of man's distrust of God's mercy.

† The latter of these two statements was thought to be Mr. Jowett's view until his explanation in the new edition of his work.

appeal to the Bible for its refutation is to beg at once the question which is in dispute, unless reasons can be supplied for believing that the doctrine of the atonement in the apostolical teaching is not one which can have received a tinge from the medium of the human minds through which the knowledge of it is transmitted.* If criticism be fairly allowed to subtract from that teaching whatever was local, or Jewish, or temporary, what tests can be suggested for showing that this hallowed mystery must not be thus surrendered? Such tests must be found in the circumstance that the atonement is not a mere belief of the Apostles' minds, but a fact of their consciousness; not the product of their logical understanding, but the reality presented to their intuitional perception.†

As this position will, however, appear to different minds to possess different degrees of strength, it is better, perhaps, to answer the suggested difficulty by some other means than the direct appeal to Scripture. The refutation must accordingly be sought in philosophy, not in theology. Restricting it to this point of view, the question will stand thus: The class of writers who suggest the argument, admit that the Scriptures convey the idea of atonement, but level against the doctrine the antecedent improbability arising from its injustice as contradictory to the moral sense. Where is the reply to be found? Is it by showing that the moral sense is simply a guide for our own conduct,

---

* Were not some of the Sermons, contained in the published course on the atonement, preached before the University of Oxford, in 1856, liable to this very charge of *petitio principii?*

† This distinction will be well understood by those who are acquainted with Reid's "Theory of Perception," or Sir W. Hamilton's remarks on it. On the application of this test to religion and inspiration, see Morrell's "Philosophy of Religion" (ch. 2, 5, 6).

not a measure of God's actions, merely regulative, not speculative?* I would yield all honor to the students who furnish such answers, and I feel a delicacy in criticising their teaching on the present occasion. Yet may it not be said, without impropriety, that those answers would not satisfy painful doubts such as those to which allusion has been made? Like all arguments which have the air of demonstration, they seem too rigorous and exact to be persuasive. They recommend themselves to the believing bystander, but not to the suffering sceptic; they constrain silence, but do not carry conviction: *ubi solitudinem faciunt, pacem appellant.* A less cogent, but perhaps more persuasive, answer is to be found, in establishing such an antecedent probability in favor of the idea of atonement

---

* This is the line which Mr. Mansel adopts in the seventh of his "Bampton Lectures," thus extending the principle of the subjective character of the faculties which he adopts from Kant to the practical as well as the speculative reason, which that philosopher declined to do. Such a view is, if true, a complete answer to the difficulties on the atonement; but would it not destroy our capacity to judge of the *evidence* of a revelation equally with its *material?* However, even if true, it is evident that it would be an answer to the opponents on new ground; the one here attempted is on their own ground.

The remarks which follow in the text are not meant to be disrespectful to Mr. Mansel, a writer who, previous to the publication of his "Bampton Lectures," had already placed himself in the highest rank of British psychologists, and who has in that work brought his knowledge of metaphysical speculation to bear on the subject of religion. Those who adopt his views, that the proper function of the moral sense is primarily directed to human duty, and, therefore, that its assertions are only presumptively true when its sphere is transferred to judge of the Divine attributes and government, will regard the line of argument adopted in this Sermon to be nothing more than a presumption drawn from the moral sense in favor of Atonement, designed to balance the counter presumption derivable from the same source against the doctrine, and will not allow to either presumption a *speculative* value.

as shall cancel the antecedent improbability which exists against it.

Whence is such antecedent probability to be drawn? From consciousness and from history; from the consciousness of guilt and from the history of sacrifice. If the idea of guilt is universal, if it contains the ineradicable conception of ill-desert, if it oppresses with incredible bitterness even those whose lives have been comparatively faultless; if also history shows the prevalence of sacrifice (no matter whether its origin be divine or human),* as instinctively suggested by the universal human consciousness as the means for the removal of guilt, and thus witnesses with inextinguishable clearness to the necessity of mediation;† then I claim that we discover here, deeply and ineffaceably

---

* See Davison on "Sacrifice;" Magee and Dr. Pye Smith on "The Atonement;" and Thomson's "Bampton Lectures," Lect. ii. The evidence in favor of the Divine origin of sacrifice is collected in Mr. Rigg's little work, before named, on "Modern Anglican Theology." The human origin of them is stated in Professor Jowett's "Essay on the Atonement" (ii, 479, 1st ed.).

† Thomson's "Bampton Lectures," Lect. ii. A remarkable passage on the consciousness of guilt existed in the first edition of Adam Smith's "Moral Sentiments," p. 204. It is quoted by Magee, "Atonement," vol. i, p. 205. The value of the argument from the idea of guilt in favor of vicarious atonement is a question of ontology, *i.e.* of that branch of metaphysical science which inquires into objective *existence* as distinct from subjective consciousness. Of course such a science cannot discover *being* as distinct from our *knowledge* of being; it can only seek to detect traces in the data of consciousness which seem to point to corresponding external realities. Assuming from psychology the existence of the idea of guilt, and the peculiarities which mark it, how far may this idea be regarded as simply indicating our own distrust of God, or how far may it be viewed as indicating a cause in Him which excites our distrust? This is the ontological problem. The boundaries of the science are stated in Mr. Mansel's article on Metaphysics, in the "Encyclopædia Britannica," eighth edition.

written in the recesses of the human heart, the conceptions which may form an antecedent probability in favor of the reality of the Scripture view, and which may cancel the antecedent objections on the other side. If so, we may thankfully acquiesce in the ordinary view of the Divine atonement, and while declining to accept unhesitatingly the trifling explanations which are usually offered for simplifying that marvel, we may cling to the mystery itself as a great reality. We pretend not to explain it, but appealing to the strength of instinctive conviction, and relying on apostolical teaching and on universal Christian consciousness as the perpetual and unanswerable proofs of its truth, we can hold fast the blessed doctrine that our Saviour's death was more important than His life, and believe that the sufferings of Jesus Christ have not merely reconciled man to God by testifying God's love, but have verily reconciled God to man by removing, in some undiscoverable manner, the obstacles, known or unknown, which prevented God from showing mercy to man. And in the light of this idea we can understand the interest which is taken in the subject in the world invisible; we can comprehend why the spirits of departed men, when they visited our Lord in that mysterious vision on the mountains of Hermon, spoke to Him of the sufferings which he should accomplish at Jerusalem.

In conclusion, let us endeavor to enforce on our consciences some lessons to be carried to our homes, and to embody in our lives.

If the sufferings of Christ be as important as we have represented them to be, we should ask ourselves whether we realize their importance in idea, and whether we attempt to live upon them in act. Each one's own conscience will tell him frankly the real state of his heart in this matter.

During the past week how often, or how seldom, have you turned in thought to the Redeemer's sufferings? The angels desire to look into these things;* departed spirits can travel back to earth and speak of them; and when, in mystic vision,† the veil which shuts out the invisible was lifted to the loved disciple, an exiled confessor, in the lonely rock of Patmos, and a glimpse, as it were, was afforded him of the heavenly world, he saw Jesus standing before the throne " as a lamb newly slain," while the choir of angelic spirits was shouting the praises, " Worthy the Lamb that was slain," and the voices of ten thousand times ten thousand of the spirits of redeemed men were uttering, louder than the sound of mighty waters, in adoring gratitude, " Thou art worthy, for thou wast slain, and hast redeemed us to God by thy blood out of every kindred, and tongue, and people, and nation." Does this subject interest heaven, and yet engages not our affections? Do we, in the hurry of our daily life, fretted with its anxieties, heated with its amusements, whirled away in its vanities, neglect to meditate on the sufferings of Christ? When we bow in prayer at morning or at evening, or kneel to receive the blessed Sacrament, do we fail to feel the value of those atoning sufferings as our only plea, and to cast ourselves on their merit for pardon, and holiness, and preparation for Heaven? Let us cling now in our daily life to that atonement, in the agony of earnest prayer, if we would wish to cling to it in the hour of death, when no other support is nigh.

And let us never forget that the souls of each one of us are dear to the Saviour. He has tasted death for every man. Though absent, He sees us and hears our prayers. Nay, in that absence He is mingling our prayers with the

---

* 1 Pet. 1 : 12.   † Rev. 5.

incense of His intercession, and carrying on the work of our salvation. You might have thought when He was absent from the disciples on the lonely mountain top He was not concerning Himself with their needs; yet in that very transaction he was engaged in a mysterious work connected with their salvation. This idea seems to have struck to such a degree the mind of that great artist whose grand picture of the Transfiguration, the noblest work in a noble gallery, formed the closing monument of his wondrous, but, alas! too short career, that he has actually violated the laws of perspective in depicting the scene. In the foreground of his picture he represents the agonized relatives of an afflicted youth imploring in vain the aid of the disciples, one of whom points to Christ as the only source of true aid. In the higher part of the picture is seen the mount of the transfiguration, and Jesus glorified before his disciples. And instead of representing the figure of our Saviour foreshortened, as it ought to have been when seen from below, he has depicted it in its full length, as if seen from the same level and close at hand.* Why was this? Was it not, think you, that the imagination of the artist, filled with that poetry, with that truthfulness, which appertain to real genius, wished to imply that the Saviour, far off on yon mountain top, to whom, as the only source of aid, the disciple was pointing, was indeed not really distant, but in truth very nigh? If he meant this, he conceived the truth. For in very deed Jesus, though far off, is very nigh to all that seek Him. Though gone on high, He is interesting himself in human salvation. He now sees each one of us, and is nigh to us. He loves each one of us

* Kugler's "Handbook of Italian Painting, by Eastlake," b. v, ch. iv, p. 384.

as He loved His disciples of old; He as much pities each one of us as He pitied the tortured beings whom He healed on earth; He as much hears and answers the secret longing, the unuttered breathing of our inmost souls, as He heard and answered the suppliants who used to petition Him face to face.

Ought we not then to flee our sins, to lay aside our half-heartedness, to yield to Him the hallowed service of a persistent will, to grieve that any portion of our hearts and our affections should be unconsecrated to Him? In the habitual practice of private prayer, in drawing nigh to His mystic sacraments, let us realize our interest in His sufferings; let us implore of Him pardon, holiness, heaven, and He will throw His everlasting arms around us while living, and put His hand under our pillow while suffering, and receive our souls into His bosom while dying. "By thine agony and bloody sweat, by thy cross and passion, good Lord deliver us."

SERMON VII.

LAWS IN THE LIFE SPIRITUAL.

(PREACHED BEFORE THE UNIVERSITY, ON ST. PAUL'S DAY, JAN. 25, 1858.)

II TIMOTHY 4 : 7.

"*I have fought a good fight, I have finished my course, I have kept the faith.*"

THERE are three aspects of human life: the life practical, the life intellectual, and the life mystical. The life practical is the lowest form of life which is strictly human, the lowest, that is, which is raised above the mere susceptibilities of sense. It may coexist with the higher lives, or it may be in great degree isolated from them. The life intellectual is a further advance. It no longer illustrates what a man does, but what he is. Its seat is in the thinking mind, as the seat of the practical life is in the active powers and conscience. There is yet a still higher life in man: the life mystical or religious; those susceptibilities, emotional and intellectual, which men experience towards the infinite, towards the unseen source of power and goodness. This form of life may perhaps be located in the exercise of a special religious feeling and in the intuition.*

These three forms of life, inasmuch as they exist as a general phenomenon, may be noticed in characters of every age and of every religion. They are facts of human nature, irrespective of the objects towards which they may be directed, and the principles under which they may be conducted. Accordingly their existence may be traced also in those persons who have embraced the Christian religion, and regulated their lives by its ideas and motives; indeed, it is in them that their highest and purest form may be studied. Though the life of every Christian must to some extent show the combination of all three lives, yet it is quite possible to select instances which shall form marked examples, more especially of some one of the three. Thus most persons who look at the characters of the Apostles of our Lord, as exhibited alike in history and in their written remains, would select St. James as the example of the Apostle who, in his exposition of Christianity, laid most emphasis on the life practical; St. Paul, on the life intellectual, and St. John, on the life mystical.†

---

* Compare Morell's " Philosophy of Religion," ch. 2. The following Sermon, in some degree, assumes that the religious life is not merely moral life elevated in its motives, and transferred to new objects, but that it depends upon a special form of emotion, which co-operates with a special form of intuition. The difference between this view and that of Schleiermacher would be mainly that he would regard the discovery of the laws of this life to be impossible. While it must be conceded that the discovery of laws in these faculties is really impossible if sought by the method of psychological analysis, a new mode for their discovery is suggested in this Sermon in the application of induction to the experiences of religious men.

† Compare this view given with proper limitations in the Rev. A. P. Stanley's very instructive " Sermons on the Apostolic Age."

And in a first view, and as a hasty generalization, there is much truth in such a statement. Broad views of this kind have their value in suggesting or directing investigation. Yet if we look more narrowly into details, we shall find that no one of these three Apostles presented these three lives in isolation. It is impossible, on the present occasion, to digress to prove this assertion of St. James and St. John; but it will be very apparent in the case of St. Paul, if we turn our thoughts in the most cursory manner to his writings. If he presents to us in the Galatians and the Romans more approach to a dogmatic view of theology than is to be found in any other inspired work; yet in the Epistles of his imprisonment, those written to Ephesus, and Philippi, and Colosse, we have the secret, inexplicable working of the spiritual mystical life alluded to, so far as language can express them. "He hath blessed us with all spiritual blessings in heavenly places in Christ." "Ye were sealed with that Holy Spirit of promise." "Your life is hid with Christ in God." "That He would grant you according to the riches of His glory, to be strengthened with might by His spirit in the inner man, that Christ may dwell in your hearts by faith, that ye being rooted and grounded in love, may be able to comprehend with all saints what is the breadth and length, and depth and height, and to know the love of Christ which passeth knowledge, that ye might be filled with all the fulness of God."* It is unnecessary to multiply passages, but we may ask, could any language, even of St. John, exceed such words as these, in expressing the depth of that spiritual inner religious life which St. Paul possessed, and of which he longed that others should partake? Nor is it

---

* Eph. 2:6; Eph. 1:13; Col. 3:3; Eph. 3:16–19.

necessary to detain you to prove, as might easily be done, that not only may the life mystical be found in the writings of this great Apostle as clearly as in those of St. John, but that also he strives to impress on his hearers the life of Christian action as flowing from Christian principle, with an earnestness not inferior even to the stern vigor of the Apostle James. The chapter on charity, in his first Epistle to Corinth, utters, as it were, the language of St. James with perhaps more than James's acuteness; and the language of St. John with more than John's pathos.

Indeed St. Paul may be adduced as an instance of an individual in whose life and teaching these three lives were very harmoniously balanced. Looking at his character as a whole, in no other Apostle can we find a model in which we can so suitably study the three in their combination in a Christian character. And perhaps it is this very circumstance which in part has largely contributed to make his influence so much more lasting and potent than that of his brother Apostles, and so operative religiously on other ages than his own. For it is observable that, be the cause what it may, the fact is real that the Apostle Paul may be measured against the first characters in history as regards the width and the permanence of his influence. It might seem a startling assertion, and yet it would bear investigation, if we were to assert to you that the single individual in all time whom we must select as having exercised the greatest influence on the world, and left the impression of his character on succeeding ages, is the Apostle Paul. If you should be at first inclined to award that proud position to some mighty conqueror, you must check yourself by the thought that the conquest has swept past like the whirlwind, and seldom left in the foundations of an improved civilization the permanent happiness which is the only compensation for the infliction of temporary misery; or if you should incline to crown with that highest honor some philosopher who has opened up new worlds of thought, and enlarged immeasurably the methods of knowledge, you must modify your decision by the reflection that such a labor, noble and enduring though it be, yields to the Apostle's work in importance, as the concerns of time yield to those of eternity. It was the Apostle who was the first to evangelize Europe. It was he who saved the Christian faith from corruption when even St. Peter himself was giving way to the pertinacity of the Jews. It was he who expounded the Christian doctrines in those Epistles which must remain the most valuable monuments of Christian literature to the end of time. And when we estimate his intellectual influence, not to take account of the many to whom he was personally known, we shall yet find that the noblest thinkers in Christendom have owed to the study of his writings those discoveries which have endeared their names. Those* who from time to time rescued from oblivion neglected truths, trained themselves by devout prayer in the study of the Apostle's writings; and if in the dark night, which in the middle ages spread its veil over the ancient civilization, there were stars showing to the pilgrim steadier and clearer light than the other luminaries of the heavens, the cause was that they reflected some rays of the Divine glory which had been concentrated in the sunlike brightness of the Apostle's inspiration. And at a later period, when the darkness of that night was disappearing before the cheering rays of the day of modern illumination, it was the study of St. Paul's writings which gave the best of the reformers those views and that courage

---

* Chrysostom, Augustin, Anselm, Aquinas.

which enabled them to break up the intellectual and religious servitude of the middle ages, and to tear away the additions which had been made to Christianity in the progress of sixteen centuries.

Further, St. Paul's example has stimulated effort as his writings have excited thought. For the Apostle stands at an illimitable distance above the most marked instances of pertinaciousness and heroic self-sacrifice. His whole heart was set, his whole life was given, to alleviate human misery, to carry the balm of sorrow to that creation which was groaning for it. Year after year he pursued his purpose, meeting no reward but the perpetual reward of conscious duty, and finding at length nothing but solitude, imprisonment, and martyrdom. Accordingly, his missionary spirit has formed the example, and has stirred the emulation, of Christendom in all moments when men have awoke to the necessity of missionary effort. How great then has been the influence of St. Paul! It was he that moulded Christianity, it was he that dispersed it over Europe, it was his thoughts that have aroused speculation and reawakened missionary effort. Judge, then, whether we are altogether wrong in claiming for him one of the most influential positions in history.

It is, not, however, to St. Paul's life of labor, nor to his mental character and dogmatic views,—not to his life practical or intellectual,—that I desire, on the present occasion, to direct your further attention. Rather I wish us to study his life as an embodiment of the life spiritual or mystical. Can we so read his life as to comprehend the growth of personal holiness in his character? Can we disentangle that which is common to all Christians from that which is peculiar to him in his capacity of an Apostle? Does he in his writings offer such explanation and interpretation of the feelings and facts of religious living as to give us the means of constructing an account of the growth of the life spiritual? To put the subject still more generally, I wish to consider the theory of the religious life, whether it is subject to laws; if so, whether we can discover them; and in that case, what the method is of such inquiry, and what the chief results to which it conducts us.

I. When we ask whether the religious life is subject to laws, the answer is so natural that the analogy of the whole of God's government, both moral and material, would suggest the presumption that the spiritual also must be directed according to a system of laws, either discoverable or inscrutable, that it might excite surprise why we should think it necessary to ask the question. Yet in reality, when we look to the history of theological opinion, we find two such very different answers given to this inquiry, that it becomes important to bestow a moment's thought upon it. There have been Christian thinkers who have said that the whole of the life spiritual is subject to merely moral laws, and have resolved the Christian life into the ordinary processes unfolded by moral psychology. This view is not common in the present age; the whole tone of thought, philosophical as well as religious, has become averse to it; but in the last century it was quite a prevalent one. Not merely critical historians, like Gibbon, who, writing from an external point of view, would naturally resolve Christian goodness into ordinary principles of moral causation, but even defenders of the faith, like Bishop Butler (we may perhaps venture to suggest), very nearly adopt the same view. What is morality, according to that eminent writer, but the restoration of a disturbed psychological constitution? and what does Christianity give us but new motives and new means towards effecting such a restoration? There is no distinct recognition of a life deeper, more hidden, kindled by the direct operation of God's Spirit in man's heart; a life consisting not merely in the restoration of a disarranged constitution, but in an actual union of the human spirit with the Divine.* It is possible that such an omission may have arisen from the controversial character of the Bishop's writings. He wished to show the reasonableness of religion, and therefore was compelled to show its naturalness. Yet the very fact that he adopted such a merely negative view in order to recommend his position may be adduced as a proof of the prevalence in his day of such a mode of viewing the question. And if it be not true of a majestic mind like that of the Bishop, it is at least true of many of the inferior Christian writers of that age, that they evince no perception of the distinctness between merely moral life, heightened and purified in its aims, and the deeper secret mystical life which Christians may possess. Among writers of this class religious changes are regarded as explicable entirely by natural laws. Circumstances or impressions acting upon us affect the feelings; the feelings form the resolutions; the resolutions produce the acts; and a course of action, connected and reinforced from time to time by the revival of the original impression through means of the law of mental association, produces at length habits, an

---

* Bishop Butler's views on this subject are to be collected from his 2d and 3d Sermon; also from the 13th and 14th, on the Love of God; and from part ii, ch. 1 of the "Analogy." In the first of these references, he shows that morality is the restoration of a disturbed equilibrium; in the second, that the love of God is the tendency, or final cause of the various emotions, if transferred to their highest object; in the third, he represents Revealed Religion as a republication of Natural Religion, together with new information, in reference to certain facts; which, however, are specially influential on men, through furnishing new motives and new means for religious improvement.

habitual course of religious goodness. Such is one answer to the question whether the religious life obeys laws. It obeys them simply because it does not transcend ordinary moral life in itself, only in its motives and tendencies.

There is another view, however, in its character the very opposite, yet which is almost equally erroneous. It is, that the religious life is a thing so mysterious, so regulated by processes incomprehensible to us, that it exists without our being conscious of it; that it is a thing which we cannot express in words, cannot think of in thoughts; that it is known to God, unknown to man; not detectable in ourselves or in others. This view has been held, in whole or in part, by many persons of different schools. It is one of the greatest of the many defects in the theology created by the Genevese reformer, Calvin. His teaching has led his followers to insist that the divine life is something depending on God's election, and not on man's freedom; that its implantation in man is a mystery; that it still exists within a man, not only when he is not conscious of it, but even,—(extravagant and inferior minds have implied this) —when he falls into actual sin. Nor is it merely among the Calvinistic Protestants that this doctrine appears. It arises also from the sacramental theory of the Church of Rome. Wherever a writer is found representing that a seed of grace has been implanted, *opere operato*, in the sacrament of baptism, which continues to exist in a man, unextinguished through years of actual sin, we have here under another form the same idea—viz., that the life spiritual is something disconnected from fact, disconnected from consciousness, disobedient to the law that religion must exclude sin; we meet here again, under another form, the notion which we have just been combating in the theology of Calvin. Yet again the same idea is found not only in

Calvinism and in the Catholic theology, but also in those Mystics who from time to time assert the existence in man (as they are pleased to term them) of faculties transcending consciousness. In our own age, through reaction against the cold, critical materialism of the last century, such a view has begun to gain ground largely. The name of Schleiermacher* will occur as the most notorious recent instance of the application of such views to orthodox theology. According to this theory there is a certain faculty in men, an intuition, which rises above all sensible objects, and penetrates into their very essence. It sees the Infinite, the Absolute, not under the ordinary limitations which sense and thought put upon the idea; but, transcending all such bounds, it scans the universe of being, it mounts to the throne of the Eternal, and sees by a supernatural intuition absolute truth, absolute goodness, absolute beauty. The life spiritual is connected with such a power. In one sense this identification may be regarded as making it amenable to laws; but in another the idea of law is thrown aside in the contemplation of it. For law is a term applicable to subordinate forms of existence and of knowledge; but inapplicable to a form of existence and of cognition which transcends the bounds of ordinary conscious criticism.

We have thus presented in opposition the two views of the life spiritual; one which would make it simply natural, the other simply supernatural. And may we not say that there is a grain of truth in both views? Spiritual knowledge is verily an apperception of truth which is not cognizable by ordinary faculties; and spiritual life is a form

---

* On Schleiermacher's Mysticism, see Vaughan's "Hours with the Mystics," book xiii.

of existence transcending even the highest moral life. Not all truth is to be reduced to that which is amenable to critical investigation. There is a world of life and of thought of which we detect the traces but cannot understand the nature. And thus far accordingly the spiritual life, be it regarded as intellectual or emotional, is supernatural; but we must be careful, on the other hand, not to disconnect the spiritual life from the human mind, nor to isolate it entirely from the ordinary facts of mental and emotional science. And it is in this respect that Bishop Butler's sermons will always have such immense value. There may possibly be in them that slight defect of an absence of any direct recognition of the life spiritual, to which I before alluded; but with this exception, we cannot study too closely the method in which he shows that even the deepest feelings, such as the love of God, are compatible with human nature.* It was the dispute which had been opened up in France, shortly before his own time, by the mysticism called Quietism,† which led him to see that such a reconciliation of the supernatural and natural was possible; and who among us does not feel what a reality it would give to many a discourse on spiritual subjects in the present day if the minds of preachers were imbued with the common sense, with the science, of the Bishop's writings?

II. In answering the first question, whether the spiritual life is subject to laws, we have almost anticipated the answer to the second, whether its laws are discoverable by man. The answer which we wish to give is, that the mode

---

* See Butler's "Sermons," xiii and xiv.

† On the history of "Quietism," see Vaughan's "Hours with the Mystics," book x.

of its operation may be understood, though its nature cannot. Supernatural in itself, it is natural in the method of its manifestation; mysterious in its origin, it is yet linked with fact and joined with psychological processes and mental laws. Under this aspect its nature will be similar to other forms of life known to us. In all such forms, whether the life organic or the life rational, there is a power and principle hidden, to which we never penetrate, but there are effects under which this mysterious principle manifests itself, of which we can write the natural history. We cannot go back to the first dawn of physical life; but beginning with its earliest manifestation in the germ cell, we can trace its manifestations onward through the various stages of embryonic and adult life to its dissolution. We can mark its features, see the action and reaction of the outer world upon it, collect its phenomena, evolve its laws, and occasionally generalize them into their causes. Similarly with the life intellectual, whether studied in the individual mind or in the history of the growth of civilization, we can mark its power, and write its history. Yet there is a residuum of mystery to which we cannot penetrate. And thus with the life spiritual; we can watch its growth, see what circumstances promote its vigor, what influences tend to blight it. We cannot know its essence; but we can write a practical history of its manifestations, and form an approximately accurate theory of its laws of operation.

If this be the case, what is the method for the discovery of those special laws?—for we have already seen that they are different from even the laws of the highest form of life, which is possible in the mere feelings when actuated by ordinary moral motives.

III. The method to be adopted for their discovery must

be to collect them inductively from the experience of religious men. There are two sources wherein we may see such experience registered: one is in the lives and thoughts of saints in the Bible; the other is in religious memoirs,—in other words, in inspired and uninspired religious biography. Facts such as those which relate to the spiritual life must be learned by consciousness alone. When we are prosecuting our researches into physical subjects, we can adopt the methods of observation and experiment as means of analyzing the facts from which we are striving to gather the inferences. But when we pass to mental phenomena these methods diminish in value; we can no longer appeal to the senses, and we are compelled accordingly to resort to the method of observation of internal phenomena which is offered through consciousness.* And you should notice that we not only trust the assertions of such consciousness when it is capable of being verified in our own experience, but we trust it also when it rests on the statements of other persons,—when it attests the mental and emotional phenomena of which other persons are conscious even when unfelt by ourselves, provided only we use proper tests to check its inaccuracy. It is by a method exactly similar to this, and by evidence similar to this, that we learn the facts which relate to the life spiritual. We must trust, even where we have not ourselves verified

---

* The methods of analyzing facts are usually stated to be three, viz., observation, experiment, and comparison; the last of which is meant to express the extended study of analogies or affinities, which is possible in botanical and zoological science. On the two former methods, see Mill's "Logic," vol. i, b. iii, ch. 7. To these methods must be added a fourth, viz, the examination of psychological phenomena through the internal observation called consciousness. This is the one to which allusion is made in the text.

them by personal experience of the like feelings, the assertions of other religious persons, on the facts of their own religious consciousness. Thus, for example, the question has often been raised, whether spiritual conversions are sudden or gradual. You must settle a question like this simply by consulting biography. There you find instances of both,—many instances like that noted one in the last century, of Col. Gardiner,* of most sudden and unexpected religious conviction;—many, again, where a good man knows not the day nor the hour of his religious change, but is only able to express his experience, when looking back to his former life of irreligion, in the words, "I know that whereas I was blind, now I see."

If we thus build our history of the laws which regulate and of the facts which accompany religious living on the consciousness of good men, what tests can we use to prevent the imposture of enthusiasm, the weakness of self-deception, the cant of fanaticism? I answer, tests similar to those which we should use in subtle facts of mental science. Thus the degree of evidence in favor of a spiritual fact would be heightened, (1) if the supposed spiritual fact be attested by the consciousness of many, and not merely of one; or (2) if it be remembered by them at a later period of life in a cooler moment, and be viewed by them exactly in the same way as when at first narrated by them; or (3) if bystanders can testify that when the fact was first asserted to exist, the persons who were conscious of it were at that time in a cool rational state of mind; or (4) if, again, the fact of internal consciousness connects itself with other facts patent to all men.† Try,

---

* See Doddridge's "Life of the Hon. Col. J. Gardiner," ch. 2.

† Some of these tests are borrowed from a note in Professor Jowett's work on "St. Paul," vol. i, p. 232 (first edition).

for example, the conversion of St. Paul by these tests. It is very evident that a miraculous conversion like St. Paul's could not admit of the first test, viz., that of comparison with the consciousness of other persons; but it admits of the rest. For it was always remembered by him, and in the main facts narrated always in the same manner (I speak, of course, here not mainly of the *outward* circumstances of the heavenly vision, in the narrative of which there are unimportant discrepancies, but of the *internal* conviction wrought in him of the truth of Christianity and of his own duty to Christ). Again, to apply the third test: he was in the situation least likely to have been affected by such a change, unless it were real; and, lastly, this internal belief of his own, that he was then touched by the Divine Spirit, connects itself with fact; for from that moment he began to preach, at great risks, the faith which he had previously persecuted.

There is yet one other test applicable to religious uninspired consciousness, viz., that of comparison with the facts of spiritual experience stated in the writings of the New Testament, and to some extent also in the older Scriptures, as in the Psalms and Prophets.*

---

* It will be observed that the view here taken of the Epistles is that they were not intended to communicate summaries of Christian doctrines, nor to be the sole standard of theological appeal; for they were written to persons already acquainted with Christianity, and instructed in the religious life by their ordinary pastors, and possessing in their own religious ideas the education necessary to explain the meaning of the Epistles. Hence the Apostle's writings were chiefly on special points, or special faults, or particular duties, and omit that kind of esoteric or detailed instruction which would be supplied to the churches by their ordinary ministers. It follows that, though all teaching which is contrary to that of the Apostles is necessarily false, yet that some truths may be held which they have omitted to name in their letters. They

If these remarks are correct, as showing the manner in which we can give the certainty of science to facts of religious experience, then we may draw from them an obvious inference, viz., that such facts may be used as a branch of the Christian Evidences. We are accustomed to regard this branch of the internal evidences derivable from personal consciousness to be a valid reason of belief in Christianity to the man who has them, but not a valid reason to be rendered to other persons, who have them not. But if such phenomena can with probability be shown to be facts, then in proportion as they cannot be explained by natural agency, they may be used as a direct proof of the interference of Heaven. Thus, for example, a religious conversion, if traces can be shown in it, after all allowance has been made for natural influences, of what appears to be Divine help, becomes immediately a *moral* miracle; and it might become a question whether the existence of such *moral* miracles, perpetuated in the Church in all ages, might not be made of similar value as an argument in favor of the supernatural truth of our religion, as the *physical* miracles which existed in the early Church.

IV. Having now seen what is the method by which we may collect the facts and learn the laws of the life spiritual, we have finally to ask whether the life of the Apostle Paul, as recorded in St. Luke's history, or as it peeps forth in his own letters, can be regarded as an instance in which we may see an embodiment of those facts and laws. It is so very obvious that it admits of being thus regarded, and most profitably also, that it is only necessary briefly to in-

---

must be truths, however, which still receive their attestation from the internal consciousness of Christian men, and not dogmas, which merely rest on tradition or on history.

dicate some very few of the points which might be gathered from it.

1. First, what facts of religious experience for all time may be learned from the Apostle's conversion? In order to answer this we must disentangle in that event the element which is peculiar to the Apostle from that which may be experienced by all Christians. His conversion appears to contain in itself three distinct circumstances: first, the miraculous manifestation; secondly, the sudden transition from Judaism to Christianity; and thirdly, the religious change from lower and mistaken views of duty to a clear perception and possession of that higher life of duty and enjoyment, the way towards the attainment of which Christ had opened up through his death. Of these three circumstances the last only is applicable as a fact of general religious history; the second was necessarily restricted to the early Christians. There was a time when, to use St. Paul's expression, they "first believed;" when they first, that is, abandoned heathenism or Judaism and accepted Christianity. Such a transition accordingly is of course impossible in countries where persons are educated as Christians. The *first* of the three circumstances, the miraculous manifestation, was peculiar to the Apostle. It seemed as if, among the disciples of our Lord, there was not one to do the work for which Saul of Tarsus was fitted. It seemed that Christianity could not go on without Saul, and therefore in that one case, and in that one only, the Saviour condescended to look down from His throne, and order a miracle, to summon an enemy to be His Apostle. "He is a chosen vessel unto me, to bear my name before the Gentiles, and kings, and the children of Israel."* Yet

---

* Acts 9 : 15.

when we have disentangled from St. Paul's conversion the miraculous ingredient, unique in his case, and the features which it shared with those of the early Christians, there still remains in it a lesson for all time, in that change which passed over him and transformed him (we cannot indeed say from sin to holiness, for he was acting on views of duty even previously, but) from the lower religious consciousness wherein he was acting wrongly, though doing it ignorantly in unbelief, to that higher religious consciousness wherein he served God as a son and Christ as a disciple. Now a change like this is, as we find from fact, one which must pass over all men, over baptized as well as unbaptized persons. Whatever may have been the benefits given under the merciful covenant of Infant Baptism (and God forbid that I should depreciate them), yet a time must come in human life when a man wakens up to a sense of his deep guilt, if he have been guilty of wilful sin, and his deep sinfulness and imperfection if even he have lived correctly. He feels that some change must pass over him before his conscience is at peace, some forgiveness be vouchsafed to him before he dare meet his God in judgment, some holiness be imparted to him before he dare hope for heaven. Such feelings and such a state of mind which, when transient, we call "contrition," or "religious impressions," and, when leading to permanent amendment, "religious conversion," may be, nay, must be, considered to have been contained, besides and in addition to all miraculous circumstances, in the conversion of St. Paul. We have indeed in his own Epistle to the Romans a clear description of this state of mind. The seventh chapter, where he describes a man wishing what is right but not having power to perform it, and crying out, "O wretched man that I am, who shall deliver me from the body of this death!" alludes to the state of a mind wakening to this struggle,—an incipient Christian. The eighth chapter, where he describes the joy and privileges of a pardoned soul, alludes to that which ought to be the ordinary state of Christian men.

2. This brings into view a second fact or law of the life spiritual, which may be gathered from St. Paul's statements about himself in his own writings. We not only find that he was changed by God's Spirit into a higher consciousness of duty, but it must also not escape notice that he always speaks of his own religious state, and implies it in relation to other Christians, with an ecstatic joy, with a confidence and a happiness, which are not possessed by many Christians now. "We joy in God," he says, "through our Lord Jesus Christ." "Being justified by faith, we have peace with God." "The Spirit itself beareth witness with our spirit that we are the children of God."* Here St. Paul speaks not merely of himself but of Christian privileges generally, and here accordingly is one of the cases wherein we may most fairly call in as evidence that appeal which I before explained, to the consciousness of Christian men. And if you consult the biographies of holy men you find that a time of their lives came when they were no longer left to infer from simple reasoning the pardon which they had received, but felt sure of it by an inexplicable peace of mind, by a direct intuition, by a consciousness of access to God in their prayers, to which they were before strangers, by an earnestness and warmth of the religious affections toward God and toward man, previously unknown to them. This is as plain a fact as any can be which is attested by the consciousness of religious men, as expressed in their memoirs

---

* Rom. 5 : 11 ; Rom. 5 : 1 ; 8 : 16.

and private writings. Along with this fact is the further one, that such a state of joy, and of love, and of holiness, was not given to them by God's election, or by some unknown mysterious process, but was simply asked and simply obtained by that one method of earnest prayer which is as sure and direct a means of obtaining spiritual blessings as any cause in nature is of producing its appropriate effects. Facts like these of religious consciousness explain the statements of the Apostle about his own joy, and peace, and love. And what may we infer from them? Not perhaps that we must disquiet the minds of those who do not possess such feelings, for their simple business is to be earnest in their religious duties, such as prayer and the holy communion, and to leave with God the result; but that we should represent such a state as attainable, that we should strive after it ourselves in prayer and in the blessed sacrament, and should proclaim it to others, for oh! let not our minds be haunted with that most fatal notion that God's spiritual gifts are capricious and various. The variation between one man and another depends mostly upon ourselves, not upon God. God has placed, if we may so say, His Divine Spirit within human power by making the gift of it conditional on, and proportionate to, man's prayers.

3. These illustrations might be immeasurably extended of the kind of use which we might make of the religious consciousness of St. Paul in discovering the facts of spiritual experience, the laws of the life mystical. But it is time to close, and we may fitly accordingly turn our thoughts to that last fact of spiritual history taught us in his life, viz., that a life of Christian duty, of Christian piety and privilege, conducts to a death of calm and

triumphant hope. If men will take care of their lives, God will not be unmindful of them in their deaths.

There is something singularly affecting in the thought of the Apostle in the solitude of his last days. At the moment when one would have thought that the Christians of Rome would have counted it their greatest privilege to be admitted to familiar intercourse with him, when converse with his mind so rich in knowledge, so mighty in experience, must have been like intercourse with an ambassador from within the veil,—he was abandoned by them all. Paul the aged—worn out with his missionary labors—deserted in his hour of trouble—abandoned in his captivity, with no one to minister to him—left without a friend as the day of his martyrdom approached—is a scene which needs neither artist to depict nor poet to describe. Yet man is never truly alone! Amid the vastness of this world's waste, or in the gloomy solitude of the prisoner's cell, God is there. The old Hebrew prophet thought himself alone when, after having traversed the sands of the desert, he hid himself in a cave among the rocks of Horeb. There, amid the mountain solitudes, Elijah was alone, surrounded only by the still scenes of unchanging nature. And yet he was made to feel that even there God was with him, that there was no such thing as solitude, that every spot through the expanse of space was present to the Almighty; that though fancying himself alone he was in contact with his Maker.

So we might be sure; indeed, we know from his last Epistle that the aged Apostle, forsaken by man, was not forgotten on high; and his dungeon—(it may possibly have been that dark, underground prison which is still shown to the traveller, which was certainly used for such purposes in the Apostle's time,* or it may more probably have been some other spot)—was tenanted by a heart filled with the peace which nothing earthly gives or can destroy, and the gloom of his prison was illumined by the presence of the God which filleth eternity, and who was as much watching over his faithful servant, and as much listening to his muttered prayers, as if this universe had been a void and Paul its sole inhabitant.

Methinks I can see the Apostle in the consciousness of this feeling kindling with rapture, as he concluded his letter to Timothy, at the thought of the mansion of his Father's house which was prepared for him; the home to him weary with the long journey of life, the haven for his spirit, long tempest-tossed in persecutions. I admire thee, Paul, in many acts of thy life! I admire thy invincible courage when proclaiming unwelcome truth before the scoffing crowd at Athens, and uttering threats of a judgment to come when standing at the tribunal of Felix,—a freeman, though in chains! But most of all do I admire thee when, the tyrant's sword being already half unsheathed, I see thee in the Roman dungeon, from which there was no prospect of emerging except to the Roman scaffold, solitary and forsaken of friends, yet looking up with confidence to that Saviour who never forsakes, and breaking forth in the exclamation, "I know whom I have believed, and am persuaded that He is able to keep that which I have committed to Him against that day."

We have not full particulars of the Apostle's death. He was doubtless beheaded; and the traveller may still visit

* It seems at least that it was used for political prisoners. There can hardly be a doubt that the lower of the two dungeons in the Mamertine prison was the scene of the execution of Jugurtha and Lentulus, and probably of the Samnite general C. Pontius.

the reputed scene of his martyrdom. About three miles to the south of Rome, on the heights which swell gradually from the eastern bank of the river Tiber, there is a solitary glen among green hills. There is the spot which is said to have been the scene of the Apostle's suffering; there was formerly, according to tradition, the Apostle's grave. "The beautiful seclusion of the region, surrounded in every quarter by the bare hilly downs of the *Campagna*, which are excavated in many spots into dens and caves of the earth, similar to those in which the early Christians so often took refuge, inspires a feeling suitable to the event."*

Though we know not the particulars of the Apostle's death, we may well picture to ourselves in imagination the calm and heroic fortitude which he would manifest as he was led out from Rome to that scene of execution. He had a conscience void of offence both toward God and toward man. If he looked backward, he had the remembrance of a life well spent; if he looked forward, he had the prospect of speedy admission into the presence of the Saviour whom through thirty years of missionary labor he had striven to love. Yet he would move forward with a deep sense of the solemnity of the moment. He is about to exchange time for eternity. In a few moments probation will have ceased with him. He will be lost or saved forever. He will be Paul the Apostle no longer; he will be simply Paul the sinner, giving account at the bar of God. Dare he trust God now,—now in this the tremendous moment of his life, now in this the crisis of his greatest need? —yes! he dare, for God is his friend. God had pity on

---

* The spot is at the Abbadia alle tre Fontane, about three miles beyond the Basilica of St. Paolo. Spalding's Italy, ii, 33. Bunsen's "Beschreibung," iii, part i, p. 460.

him when he was yet a sinner; surely, therefore, He will have pity on him now that he has been His servant. As these thoughts pass through his mind, the sadness passes from his face, a look of seraphic joy spreads over his countenance and illumines every feature; and as he yields up his neck to the sword of the executioner, he exclaims in triumphant exultation, "I am now ready to be offered, and the time of my departure is at hand. I have fought a good fight, I have finished my course, I have kept the faith. Henceforth there is laid up for me a crown of righteousness, which the Lord, the righteous judge, shall give me at that day; and not to me only, but unto all them also that love his appearing."

SERMON VIII.

THE GIFTS OF THE HOLY GHOST.

(PREACHED AT THE CHAPEL ROYAL, WHITEHALL, ON WHIT-SUNDAY, 1858.)

JOHN 14:16.

"And I will pray the Father, and he shall give you another Comforter, that he may abide with you forever."

It was a sad and anxious moment to the Apostles when they stood on the top of the Mount of Olives and saw Jesus parted from them and a cloud receive Him out of their sight. We may reasonably imagine that they stood gazing into heaven, doubting whether Jesus had ascended from them forever, or whether His departure was only one of those many mysterious disappearances which they had witnessed in the forty days which had succeeded to His resurrection, when the heavens had suddenly yielded up to them His presence, and Jesus had stood in the midst, and had as suddenly vanished out of their sight. They might well think that he had only gone away for a season; but these hopes were dissipated by the appearance of the two heavenly messengers, who assured them that Jesus had taken a final farewell, and had departed till the last great day: "Ye men of Galilee, why stand ye gazing up into heaven? This same Jesus which is taken up from you into heaven shall so come in like manner as ye have seen Him go into heaven." It was then that the Apostles first felt their loss. It was then that they knew their loneliness. A band of men—most of them rude fishermen from a northern province—had left their occupation to follow a wonderful teacher, as his associates in subduing the world; and now he had vanished and abandoned them to subdue that world, as it seemed, unaided.*

It was at a moment when such thoughts as these filled the Apostles' minds, that they would begin to turn their hopes to that mysterious promise which their Master had not long before given them of a Comforter who should be with them in his absence. "And I will pray the Father, and He shall give you another Comforter, that He may abide with you forever: even the Spirit of truth. I will not leave you comfortless; I will come to you. Yet a little while and the world seeth me no more; but ye see me; because I live ye shall live also." Nor did they wait long in doubt, for after about ten days the surprising miracle happened, that there came from heaven a sound as of a mighty rushing wind, and they were all filled with the Holy Ghost, and began to speak with other tongues as the Spirit gave them utterance. Of all miracles ever wrought, the gift of the Spirit was the most astounding.

That marvel is a surprising one which the Scripture

---

* Their feelings of solitude at that moment have been beautifully depicted by a Spanish lyric poet, Luis de Leon, in his hymn, "En la Ascension," beginning "Y dexas, Pastor Santo," &c., a translation of which, hardly inferior to the original, will be found in Ticknor's "Hist. of Spanish Literature," vol. ii, ch. 9: "And dost thou, holy Shepherd, leave," &c.

opens up to us in the miracle of creation, when it places us at the dawn of created nature; when it transports us backward to the depths of a past eternity when God was alone. Then as now God was; but besides Him there was nothing. The Supreme Being existed with universal silence round Him. Suddenly his fiat went forth, and the universe was peopled with motion and life. Orbs began to roll in periodic circle round his eternal throne, and intelligences, sparklings of the Infinite, sprung into existence at His bidding. "The morning stars sung together, and all the sons of God shouted for joy." That, again, was a stupendous miracle worthy of a God, the loftiest expression of the tenderness of the Almighty, when his own Son was born in the village of Bethlehem; stooping to join mankind in their sufferings, that he might elevate human nature along with Him to the throne which he had left. Well might the choir of the Heavenly hosts break in upon the stillness of the midnight with their chant of triumph! Well might inanimate nature respond to the event by launching forth a meteor to attract the Eastern sages!

But majestic as was the miraculous sight of the freshness of the morning of created nature, stupendous as was the condescension in God becoming man, the miracle was, if possible, still more marvellous when God the Spirit condescended to come down to take His residence in the hearts of men. It was an infinite condescension for God to live among men, it was a greater one for Him to make His dwelling-place within men's hearts. It was a wondrous comfort for His disciples to be able to go to a God present on this earth and ask His aid; but it was a mightier privilege to know that, without undertaking a long pilgrimage to seek the presence of a local Saviour, there was help to be found from an omnipresent Comforter; that for men of every race and rank, without respect of age or sex or condition, for the captive and for the free, for the sick and for the strong, there was close at hand a Spirit to be given in answer to their prayers; that wheresoever under the broad heaven, on earth, or on sea, in the crowded city or in the solitude of the desert, one prayer is breathed up to God for His Spirit, then from the Invisible that Spirit breaks forth; and though not with rushing wind or tongues of fire, yet within the soul, heart to heart, spirit to spirit, He witnesses by the gift of conscious comfort and purifying holiness that He is present, the Comforter, in Christ's absence, who shall abide with men forever.

We cannot wonder that at the first manifestation of such descent His presence was marked by evidences such as the world had never before seen. It is not surprising that the overpowering joy of the Spirit's presence in the hearts of that band of waiting worshippers caused it to be supposed that they were drunk with wine. We cannot wonder that the greatest of events should be marked with the greatest of effects, that ignorant men should suddenly be strengthened to speak in other tongues; that the stammering disciple should be suddenly turned into the eloquent preacher, and the fierce and prejudiced Jew into the loving, subdued, hallowed Apostle.

What, then, was the special nature of that outpouring of the Spirit? what the special gifts which His presence communicated? This is the practical and important question for us, if we would ascertain how far we have become, or may become, partakers of them. They were especially four: miracle, inspiration, holiness, and religious usefulness. We assert, first, that the gift of the Spirit conferred the power to suspend Nature's laws by the action of what is usually called miracle; secondly, that it strengthened the human intelligence to penetrate the world of spirit, and gaze face to face on undiscovered truth and reveal it to mankind; thirdly, that it was the means of changing unholy men into holy ones; fourthly, that it was manifested in a mode which, for want of a better term, may be called religious usefulness,* by which we mean that it accompanied the ministry of those who possessed it in such a manner that their words produced, in a supernatural degree, a moral effect on those that heard them. But while we assert that these four gifts were conferred on the Apostles by the miracle of Pentecost, we do not claim them all as the privilege of ordinary Christians, nor as the permanent gifts which the Divine presence was to confer. If we briefly survey each class, we shall be able to understand how many of these gifts were temporary and how many perpetual; which of them were special and which general.

1. The first form in which the gift of the Holy Spirit manifested itself was in conferring the power of working miracles to aid in the propagation of Christianity.

There is scarcely a fact in the history of the world better attested than the Christian miracles. The whole evidence of history must be belied if their existence is denied. The proof by which the reality of the great characters of history, and of their deeds, is established, is hardly more certain than the evidence from contemporary testimony of the existence of the Apostles, and the actual reality of their miracles.†
And it is a proof of the cogent evidence which has been

---

* The meaning of this phrase will be made clear hereafter. It is intended to be comprehended (as well as prediction of the future) in the apostolic word "prophecy," in 1 Cor. 12: 10.

† The clever work of Archbishop Whately, "Historic Doubts on Napoleon Bonaparte," has for its object the establishment of this point.

brought to bear upon this subject, that sceptics have latterly ceased to attack it.* There may be difficulties in religion, and there may be weak points in the evidence of it, but this weakness is not in the department of miracles. If indeed we had no contemporary proof, we might almost argue that some kind of supernatural support was vouchsafed to the early Christian missionaries, if we only measure their success against the means employed.† A few illiterate peasants from the shores of the lakes of Galilee set forth to convert the world. As they spoke, men were pricked to the heart, and joined themselves to them. By the testimony of the heathen philosopher Pliny,‡ the converts became changed men, and the progress of the religion was so great, that whole provinces in a short time bowed before the new faith. Was this a common religion like the many systems which this world has seen? It bore this difference: it was a system narrow, exclusive, unadapted to the tastes of men, though marvellously adapted to their moral wants. It was favored by no earthly power, it offered no earthly reward, it pandered to no prejudice, stooped to no passion, admitted no collusion. What quality then did it contain which proved so attractive? Search among the religions of the world, and you will perceive the contrast which it bore to them. Try to penetrate through the mists of twenty-four centuries of Hindoo history to the origin of Boodhism, and you will find that it was not a new

---

* The allusion here is to the fact that modern critics (*e. g.* Strauss, "Leben Jesu") have found it necessary to explain away the Evangelists' narratives by reducing them to myths instead of denouncing the miracles as simple fables, as writers of the last century were accustomed to regard them.

† See Milman's "Bampton Lectures," Lect. vi, p. 269.

‡ Pliny's "Letter to Trajan," Ep. x, 97.

creed, but the vigorous expression of a moral reformer who awoke the religious instincts of his countrymen.* Trace the rise and progress of Mahometanism, and you will discover that it owed its progress to the sword, or to the lust of war, or that it was at best but the reflection of Arabian thought, the embodiment of the primeval patriarchal unitarianism, which had always swayed the thoughts of those sons of the desert. Christianity, on the other hand, aimed a blow at every prejudice, and was founded on a revulsion of previously-known principles. How can you account for the mightiness of that result from the smallest of causes? Why did the heathen world bow before the messengers who came forth from a retired district of an insignificant Roman province? Suppose a person meditating upon this circumstance without being acquainted with the facts of the history, would he not be compelled to admit some potency of earthly evidence, or some proof of supernatural power?† And that presumption is true. Christianity made all its advances by argument and proof. Its missionaries healed the sick and raised the dead, and miraculously spoke

---

* The view of Boodhism here adopted is that which would make it part of a movement in the East to which Confucius's revolution in China is analogous, probably about 600 B.C. But see the account of the Journeys of the Boodhist Pilgrim, Hhouenthsang, recently translated from the Chinese.

† Most readers would admit that Gibbon's enumeration of the supposed causes which may have led to the spread of Christianity is utterly inadequate to the explanation of the phenomenon. It is true that the five causes which he suggests really were not without their influence, for Christianity was so all-embracing in its operation that it included those among others; but they were *conditions*, not *causes*, of its spread. Gibbon's chapters on Christianity are rather a subject of regret than of alarm or bitterness. It is pitiable to see a mind such as his, unapproachable in its greatness,—the Michael Angelo of history—demean itself by such sophistry.

tongues which they had never learned; and men saw these things and marvelled, and perceived that God had spoken, and they gave ear and accepted the doctrine which those mighty wonders attested. We pause not to inquire into the possibility of miracles; we cannot even spare a moment, on the present occasion, to reconcile their existence with the great government of the Almighty by general laws; but we accept the fact, and we refer to the mighty outpouring of the Spirit at Pentecost as the cause; and we recognize in the potency of these evidences the first of those great acts of help to the weakness of the early missionaries, which led Christ so emphatically to describe His offices, when he predicted his coming, as being those of the Comforter.

Was then this gift of miracles temporary or perpetual, special or universal? The answer to this question must be found in the fact. Miracles have ceased, and hence we argue the temporary character of it. Antecedently we should have expected its perpetuity; but we accept the experience of history as proof of the error of our anticipations. It is unnecessary to pause to ascertain whether miracles have been occasionally repeated at long intervals by the pious faith and prayers of saintly men.* We rather would ask what was the reason why the gift, once so abundant, was so soon withdrawn. The reason would appear to be this. The office of miracles was merely to awaken human curiosity to the heaven-sent message; when men once turned their ear to listen, the doctrine was allowed to speak for itself. The perpetual evidence of Christianity, unassailable by the advance of science or the acuteness of criticism, is that it links itself to every human want and

---

* Compare Dr. Arnold's "Lect. on Mod. Hist." p. 105.

responds to every human susceptibility; and thus when the missionaries had once established by miracles their claim to be considered Divine messengers, the mighty proof was withdrawn, and Christianity was left to work its way by moral evidence. The external demonstrative proof was only for the purpose of arousing attention;* the internal appeal to human candor was to effect the rest. The Almighty was not in the earthquake nor in the storm, but in the still small voice of persuasion and of conscience; and so we think we can discern it to be a beautiful example of Almighty wisdom that miracles were withdrawn when they were no longer needed, that the Spirit ceased thus to comfort when the Church was no longer mourning for this aid.

2. We asserted, secondly, that the influence of the Holy Spirit strengthened the minds of the Apostles to discover truth and elevated them to reveal it with an inspired authority. This was a blessing which Christ had distinctly promised to His disciples when He said, "I have yet many things to say unto you, but ye cannot bear them now. Howbeit when He, the Spirit of Truth, is come, He will guide you into all truth, and He will show you things to

---

* The comparative logical weight of the internal and external evidences of Christianity is a subject which still demands treatment. The view here intended is that the external evidences, such as miracle, prophecy, martyrdom, &c., were mainly designed to arrest attention, in order that the internal appeal to the moral consciences of men might have a fair field for producing its appropriate effect. It is observable, in confirmation of this view, that on the only occasion on which we find St. Paul to have met with an intellectual audience, able to appreciate evidence, and willing to listen to it, viz., at Athens, he wrought no miracle; but appealed wholly to argument. If this view be correct, political power, in obtaining a hearing for Christianity in heathen lands, performs at this present time the analogous function to that which the external evidences fulfilled in the early ages of the Church.

come." Our Lord had taught a germ of truth while on earth,* but He here promised to send to the Apostles a Divine Spirit which should illuminate them to discover a deep and mystical meaning where they had not before seen it, and to penetrate into the world of unseen realities, and

---

* The view here intended is that the teaching of the Apostles, after the Pentecostal gifts, was in advance of that which our blessed Lord communicated while on earth. This can not only be shown by fact, but is implied by the Evangelist when he said that "the Spirit was not then given, for Jesus was not yet glorified." (John 7 : 39.) The work of our Lord on earth as a *teacher* was the reformation of doctrine and practice; that of the Apostles was reconstruction. Our Lord taught the spiritual meaning of the Jewish law, and showed that Judaism was fulfilled; the Apostles taught that Christianity was not merely Judaism fulfilled, but Judaism abolished. Antecedently to the Transfiguration, our Lord taught only that he was the Messiah; subsequently to that event, that he was the Messiah *to suffer;* but it was the Apostles, and specially the writer of the Epistle to the Hebrews, who were the first to explain the doctrine of the Atonement, the interpretation of the Jewish law, and the internal spiritual life of Christians. According to the view here advocated, the stages of Divine revelation would be as follows, each in advance of the other: (1) the Patriarchal; (2) the Mosaic; (3) the Prophetic; (4) the teaching of St. John the Baptist and of our Blessed Lord until the Transfiguration; (5) the teaching of our Lord after that event; (6) the teaching of the Apostles. The difference between the Mosaic and Prophetic dispensations is made clear by Davison in his "Lectures on Prophecy." The growth in the teaching of the Apostles is made clear in Neander's "History of the Planting of the Early Church." Evidences of smaller differences and of less plainly marked advances might be found under most of these periods. Thus, for example, in the last of those mentioned, the Apostolic teaching, we might perhaps enumerate as subordinate distinctions: (a) the Jewish school of teaching of St. James and St. Peter; (β) the Gentile school of St. Paul; (γ) the Alexandrian school of the writer of the Hebrews; and (δ) the intuitional school of St. John. "But all these worketh that one and the selfsame Spirit, dividing to every man severally as He will."

evoke from its depths discoveries new in character and invaluable in import. It must indeed be frankly admitted that a human element, as well as a divine, is traceable in the writings of the Apostles. Those favored men did not lose their personality beneath the overpowering majesty of the mysterious inspiration. They were not mere automatons, mechanically uttering words which they understood not. Yet, while they were left to express the divine truth according to their different habits of mental thought and different modes of human expression, there was in the truth which they conveyed a great reality which they had not discovered by unassisted reason,—a reality into which the Spirit of truth himself had deigned to guide them. The treasure was in earthen vessels; but in itself it was divine.

The question will suggest itself whether this great gift of the Comforter remains, or whether it, like the gift of miracles, has departed. The answer must be given that in great part it has disappeared, and for a similar reason. As the gift of miracles was continued only so long as was necessary for gaining a hearing for Christianity among the heathen, so the gift of inspiration was continued only so long as was necessary for evolving the body of Christian doctrine. Yet there is a sense, though a far humbler one, in which the Spirit does still illuminate religious men. Such is the play of human emotion with human thought, and especially when the mind attempts to judge on religious questions, that it is impossible for a man (be his clearness of mind as great as it may) to apprehend religious truths unless his heart be touched to feel the value of that truth, and his prejudices lulled to allow him to give it a hearing. It is this great gift which the all-sanctifying Spirit confers now as of old. We do not claim that He interferes with the mental laws which govern thought, but we claim it as a fact alike of human experience and of psychological science, that some means of controlling human prejudice, in order to secure an honest judgment on such questions, is necessary; and experience seems to prove that the Spirit of God vouchsafes such help even now to all that seek it, and will, we may infer, continue to do so as long as that help shall be needed, *i. e.*, until time shall be no longer.

3. The consideration of the last-named gift of the Spirit leads, by a natural transition, to that which was noted as the third of His blessed offices, viz., the changing unholy men into holy ones. The manifestation of this great fact is seen in the history of the Pentecostal miracle. The intellectual growth of the Apostles' minds by that visitation is not more marked than the moral change which passed over their characters. It is not more extraordinary to find the blind prejudices disappear from the Apostles' minds, and the ambition for an earthly monarchy fade before the clearly seen vision of the spiritual kingdom, than to see the weak made strong, the timid bold, the fierce hallowed into meekness.* The selfsame Peter who had quailed before the suspicion of the servant-maid and had denied with oaths his knowledge of Jesus, now stood up before assembled crowds to proclaim the crucified, with a courage which Christ had foreseen when he had declared that he was the rock on which He would build His church.† The same John who had wished to call down fire on the Samaritans for their incivility now became a

---

* See Neander's "History of the Planting of the Early Church," vol. i, ch. 1.

† For the explanation of this promise and its fulfilment, see Rev. A. P. Stanley's "Sermons and Essays on the Apostolic Age."

pattern of gentleness; the Son of Thunder was subdued into the Apostle of Love. Stephen proved himself to be so influenced by this mighty change that in a few days he yielded up his spirit with heroic fortitude in triumph, muttering with his dying breath forgiveness towards his murderers. And not in the pillars of the Church only, but in the humble members of it, there is evidence of the same absence of selfishness, the same devotion to God, and the same love of man.

Let us pause a moment here to contemplate this wondrous work; for we are not here dealing with a gift which, like that of miracles or inspiration, has passed away. This is the everlasting gift which the Holy Spirit confers as really, as fully, in the present time as formerly. Now as then, He changes unholy men into holy ones. Now as then, He works in the hearts of men that mighty change which is the indispensable preparation for admission into the presence of the God who cannot bear sinfulness.

We possess a nature prone to evil, and the whole system of our moral affections is, in a considerable degree, disarranged and disorganized. We love sin, we do not love God; we are fettered by the bond of selfishness; our generous instincts are repressed; though endowed with capacities which no finite object can satiate, and made to strive after the infinite, we abdicate these lofty aspirings, and allow ourselves to be absorbed by the present, and our eyes grow dim to the eternal and the future. It is the office of the Spirit of Christ as the Comforter to remove this evil. Acting on a man through the instrumentality of conscience, the Spirit rouses him from the lethargy of his nature, and excites in him apprehensions of God's hatred of sin and of a judgment to come; sometimes drawing by motives of fear, sometimes by motives of love. And when the individual is convinced of sin and is sensible of his miserable condition, the Spirit suffers him not to be overwhelmed by the sight of his own worthlessness, but encourages him to seek for mercy from him who is mighty to save. Those influences smite only to heal, they awaken the sense of dependence on God's mercy in Christ, they incline the man who is the subject of them to prayer, they stir up within him inexpressible longings after holiness and goodness. Thus by a gradual or sometimes a rapid progress, of which we are not permitted to trace every stage, the heart of stone is removed, and the heart of flesh is substituted. The soul which once loved sin begins to love God, the selfishness which once ruled dies away, and the generous instincts of love struggle, as with the force of a pent-up fire, to express themselves in acts of mercy. We cannot now pause to trace the continued progress of goodness in that soul; the help which is vouchsafed to it in its sorrows, the support in its temptations, the grace communicated to it in the sacraments; yet as we mark that mighty change, and think of the Spirit dwelling in that heart, and remember that this gift is for us and for our children's children as much as for the churchmen of old, can we fail to understand how truly our Lord called the Holy Ghost a Comforter, when He promised, "I will pray the Father and He shall give you another Comforter that He may abide with you forever?"

4. But the catalogue of blessings imparted by the Comforter, and implied in the promise of His gift, is not yet complete. We asserted that, besides the gifts of miracle, and inspiration, and holiness, there was a fourth conferred in Apostolic times, the gift of religious usefulness, by which was meant that the Spirit of God miraculously and mysteriously accompanied the words of the Apostles to the

hearts of men. It is this gift which is called in St. Paul's Epistles, in his catalogue of the gifts of the Holy Spirit, the gift of "prophesying;" which did not (as it would seem) imply solely the power to predict events, but was a special power, analogous to the gift of natural eloquence, of expounding religious truth with probably a preternatural power of operating by means of it on the hearers. Here again we encounter a gift which mainly was confined to the Apostolic age; yet, surely, not wholly so. For it is not, surely, wrong to anticipate Christ's presence to accompany His word and His sacraments to the end of time; nor is it mere hypothesis to suppose that, whenever in the history of the Church there has been a great religious awakening, there has been a manifestation of the operation of the Spirit of God on men's hearts; that wherever an apostolic man has arisen, burning with apostolic love, and moved with apostolic zeal, and praying with apostolic piety, and exercising a ministry marked by apostolic success, the Church has seen repeated in such earnestness the Holy Spirit's gift of prophesying, the evidence of the continued operation of the Comforter. Proofs of this assertion cannot be offered in the limits of this discourse, yet an allusion may be made to one or two examples of the great outpourings of religious influences which history has presented, as evidences that the moral power of Christ's Spirit was not confined to the Apostolic age; and the examples shall not be selected from the catalogue of those movements to which the world is accustomed to appeal as the visible evidence of the operation of God's Spirit in dissipating error and improving civilization; but shall be drawn from among the mystics of the earth, from men whose primary object was not the advance of civilization, and in some of whom the heavenly truth may not have been unmixed with human error; for in them the effects of goodness which are traceable, will for that very reason be more naturally ascribed to superhuman energy.

The first instance shall be purposely selected from one of the apparently fanatical movements of the middle ages, and from the history of a Church from which the established Church of England has justly separated itself, and against the doctrines of which it specially protests; a Church in which, nevertheless, God in His mercy has been pleased to stir up souls for Himself in spite of the manifold errors which have impeded their holy work, and dimmed their perfect brightness. For it is one of the special glories of this hallowed festival,* that in commemorating the gifts of the universal Spirit, we can outstep the narrow limits of creed or party, and thankfully trace the Spirit's work in all who, in every place or in every creed, have feared God and wrought righteousness.

About the year A.D. 1200,† in a retired town of Central Italy, called Assisi, there lived a gay young man who was brought to the gates of the grave by illness. He felt death at hand, and was unprepared for it. The Spirit of God touched his heart; he began to prepare himself for it; and when health unexpectedly returned to him, he threw aside his gaiety, and retired to the mountains to lead the life of a hermit. There in the solitudes of the grand chain of mountains which stretches through Central Italy he communed with God, as the prophets of old; and returning in the strength of his pious convictions, he spent his life in

---

* Whit-Sunday.

† The writer, since preaching the above Sermon, has had some doubts whether the religious influence of Francis of Assisi has not been exaggerated. The best estimate of his character is to be found in one of Sir J. Stephen's Essays on "Religious Biography."

arousing men to a religious life. The plain which surrounded his humble dwelling was frequently occupied by thousands of weeping penitents. His followers formed themselves into one of the monastic orders, which has always made itself remarkable above its rivals for its devotion to the wants of the sick and the poor; and after leading a saintly life, the founder died, and has been ever since held in reverence throughout Christendom under the name of St. Francis of Assisi. I will yield to no one in undying attachment to the Protestant faith, yet I envy not the heart of that traveller who can visit the sanctuary which covers the saint's grave without acknowledging the work of God's Spirit in his life of piety. That sanctuary is a temple of art.* Its walls are adorned by frescos, from the hand of the great masters who revived Italian painting. Yet far more glorious than the physical glories of that wondrous landscape of the southern clime which surrounds that temple, more glorious than the works of human genius traced on its walls, is the remarkable work of the righteous man whose body lies there enshrined; and I should think badly of the piety of the pilgrim who could bend over the saint's grave without a tear, and who could turn away without a prayer that God would be pleased to grant him, free from error, a small measure of that heavenly love which burnt so brightly in the spirit, and marked the life of Francis of Assisi.

We may cite another instance of the merciful influence of the Spirit of God in religious revivals which were almost similarly erratic, though not corrupted with a like admixture of superstition, in times nearer to our own, and in our

---

* An account of these frescos may be seen in Sir C. Eastlake's edition of Kugler's "Hand-book of Painting," vol. i, b. iii, ch. 1.

own land. The religious state of England was never, since the Reformation, brought to so low an ebb as at the commencement of the last century. The civil wars of the preceding age, and the licentious influence of the court and of the literature of the time of Charles II,* had led to the deterioration of the national character and the almost total extinction of religious life. What were the means and who were the instruments through which the flame of earnestness was rekindled? We should belie the facts of history if we were to deny their due meed of praise in this revival to those individuals who founded, in the last century, the irregular religious systems then common under different names.† A few clergymen of the University of Oxford,‡ assisted by many laymen, went about our land preaching with the earnestness of the friars of the middle ages, exciting attention by the very eccentricity of their movements, and arousing the religious feeling of masses of the population; and though most will regret that the earnestness was lost to our Church, yet all must acknowledge its beneficial influence in removing the cold lethargic state of feel-

---

* The allusion here is specially to the dramatic literature of that age, an estimate of which is given in Hallam's "Hist. of Lit.," iii, ch. 6, and in Macaulay's essay on "The Dramatists of the Restoration." The drama may be regarded as both a cause in the formation of a nation's character, and the index of its moral tastes.

† The name of Methodism is the best known of these systems; but there were several other (though less important) centres of religious influence at the same period. Several such appear in the "Memoirs of Selina, Countess of Huntingdon," and in "The Life of the Rev. G. Whitfield."

‡ The Wesleys and their early friends. See Southey's or Watson's "Life of Wesley;" or "Rev. J. Wesley's Journals," vol. i; or a work on the history of Methodism, entitled "The Centenary."

ing which previously existed, and in carrying the light of truth into many a benighted region of our land, neglected in real heathenism.

There were agencies too within the Church itself, which Providence set in operation, for stirring up the hearts of men. One individual pre-eminently, in the University of Cambridge,* through many years used his position to instil into the minds of the students the necessity of an earnest practical personal piety. You may not accept theology exactly as it was there presented. You may think that system to have been a very narrow one, and very unscientific. You may lay more stress upon the Sacraments, and less upon election; more stress upon prayer and less upon faith; yet venture not to deny the vitalizing influence of the Spirit of God through that teaching, in awakening the Church to the present energy which stirs the hearts of men of all parties. We may vary in opinion from those individuals that we have named, yet I cannot but express my belief that if, in centuries to come, some future Neander† should attempt to gather up the memorials of piety and of earnest efforts which marked the eighteenth century, as that distinguished historian, who lately was removed to the Church triumphant, collected those which existed in the earlier ages of the Church, he will find many of his brightest examples, many of his enduring confessors, in those who have received the faith and admired the labors of Wesley, of Whitfield, and of Simeon.

We have sketched the four principal blessings which the great gift of the Holy Spirit conferred on Christ's Church, —miracle, inspiration, holiness, and usefulness,—and have

---

* Rev. C. Simeon.

† Neander, as is well known, devoted great attention in his history of the Church to the study of the internal spiritual life of Christendom.

noticed which were temporary and which are perpetual. The great gift of holiness is for us and for our children forever, and to as many as the Lord our God shall call. I wish, in conclusion, to impress upon you the necessity of examining yourselves to ascertain whether you have the Spirit's gift of holiness; and if not, to urge upon you the necessity of seeking it. If you have no consciousness of your own sinfulness, of your exceeding great needs; if you never know what it is to drop the tear of penitence, you have yet to take the very first steps in conscious religious living. And to those of you who have in some measure set out in a real effort towards a religious life, let this subject be a warning to see how far, how very far, you fall behind that standard of holiness, and happiness, and usefulness to which you might attain. It rests with yourselves to attain that bliss. The Apostles were ordered to continue in prayer to obtain the Comforter; and the law is now as then. They that ask receive. In religion, as in common life (we say it without irreverence), God helps them that help themselves. Make your earnest and constant supplications to God for the blessings of pardon, of holiness, of consolation, and they will be given. The Comforter has been imparted, and God has been pleased to place His influences within human reach by making them to be obtainable by prayer.

SERMON IX.

PROVIDENCE IN POLITICAL REVOLUTIONS.*

(PREACHED BEFORE THE UNIVERSITY, JANUARY 30, 1855.)

PROVERBS 16 : 4.

"*The Lord hath made all things for himself; yea, even the wicked for the day of evil.*"

THE Hebrew monarch here expresses his conviction that the whole course of nature and of history is superintended by the providence of God, so that even the plans of wicked men are made mysteriously to co-operate in carrying out the divine purposes. The thought is indeed expressed in the form common to many passages of Scripture in which God's providence is so spoken of, as to seem to exclude human freedom. Thus the Lord is said to have " hardened Pharaoh's heart," to have " put a lying spirit in the mouth of Ahab's prophets;" wicked men are stated to be " fore-ordained to condemnation," and the righteous to be " pre-

---

* The abolition of King Charles's day, and of the other political services, seems to render an apology necessary for the publication of a Sermon on a subject now so obsolete. As the publication of it was,

destinated to eternal life;"* all which texts must be understood only to refer to that general scheme of divine government which sees the end from the beginning, and in no sense to exclude the idea of man's free agency and responsibility.

The reason of such a mode of speaking may perhaps be found in the tendency which history shows to have always pervaded the Oriental mind, of looking at the acts of men from the divine side to the exclusion of the human,—a tendency which, in its ultimate development, has degenerated into the fatalism of Eastern creeds; and herein we may observe one of those instances where God, in His great gift of a revelation, has made use of the peculiarities of human thought as the medium of its transmission. The treasure is divine, but it is communicated in earthen vessels. The theology of the Bible is inspired; the ideas which it offers of God's government and of man's character are realities; but the form under which those ideas have been conveyed has partaken of the peculiarities of personal or of national thought which belonged to those who uttered them, and indeed has been accommodated with a marvellous wisdom to the circumstances and wants of the ages and peoples to whom they have been addressed; and it is for this reason that, in passages such as our text, in which the wicked are said to be "made for the day of evil," we understand the inspired thought (when translated into the modes of thinking of European nations) merely to be that the universe of events is so arranged and overlooked by the Father mind that even the day of evil, which wicked men,

however, earnestly requested at the time when it was preached, and as the subject is treated in such a manner as to elicit political and historical principles which can never become extinct, the writer ventures to allow the Sermon to appear in print.

* Exod. 14 : 4–8 ; 2 Chron. 18 : 21 ; Jude 1 : 4 ; Rom. 8 : 29, 30.

acting in the strength of human liberty, bring about, is made to adjust itself harmoniously—a wheel, as it were, within a wheel—into the vast scheme of nature; and so, ultimately, while the wicked bear the burden of their own personal responsibility, to work out the great purposes which an all-good God may design in the government of His works.

If the thought of Solomon be taken in this sense, it will be easily seen how many circumstances had occurred, both in the history of his family and his nation, to bring home to him the conviction of this truth. A mind like his must, doubtless, have often marvelled as it meditated on the remarkable deliverance of his nation from the land of Egypt; for the light of prophecy had shown that this whole passage was no accident in the national history, but that the evil had contributed to bring out the result equally with the good; or if he thought of that period of four hundred years, which intervened between Moses and the Prophets, during which there was no open vision, and in which the Almighty might almost seem to have left the nation to the superintendence of merely ordinary laws (as occurred afterwards in the corresponding period between the Prophets and the Gospel), he may perhaps have marvelled how wondrously, in spite of deep internal disorganization and occasional anarchy, the successive schemes of the border nations to annihilate the chosen people had not only been defeated but made to minister to its good; and if he passed on in thought to the circumstances which had attended alike the early and the later life of his father David, how many a day of evil, both of public and family history, would seem to his pious mind to have been ordered of God! how frequently would events be seen to have tended to the very opposite effects to that for which their wicked authors had designed them, just as the waters of a river steadily press onward to the ocean, even when, by the windings of the course, they appear to be flowing directly away from it! how often in such meditations as these, might the heart of Solomon, overflowing with gratitude for the past and exulting with confidence for the future, express its experience in the words, "The Lord hath made all things for himself; yea, even the wicked for the day of evil."

But if the experience which Solomon had of the world's history, restricted as it was to that of a single nation and of a short period of time, yet brought home forcibly the conviction of this law of Providence, how much greater opportunity has been afforded to us of noticing the evidence of its truth. In his day the drama of the world's history had hardly begun to be acted; it was not merely, as it were, in its first act but in its first scene, and it would have defied the skill of the acutest to have anticipated its development; and though the catastrophe is not yet come, yet we are able to study the plan of Providence in several distinct epochs and in many distinctly marked manifestations under each. We can pass in thought beyond the Jewish nation, and view the successive centres of power and of civilization which sprung up in the ancient world. We can study the display of human passion in its fiercer and darker forms, in that period of general convulsion, of the universal extinction, as it at the time appeared, of all order, and maturity, and goodness, which marked the overthrow of the great empire which had absorbed the other powers. We can see how the darkness of that night of barbarism passed away before the voice of Him who commanded, Let there be light; and how the brightness of law, and learning, and liberty dawned again upon the earth. And still further, in each of those empires which have arisen in the modern world, many instances, alike in their external and internal history, afford proofs of the Almighty power, of the Providence, of Him who sitteth above the waterfloods and calms the storm of human passion. And as the pious mind watches in all these successive periods the agency and operations of evil, and the marvellous manner in which, on the whole, the good has resulted, he must feel that the text is brought home to him with a fulness of proof which Solomon could not possess. The voice of history is seen to declare, " The Lord hath made all things for himself; yea, even the wicked for the day of evil."

The service of this day naturally directs our thoughts to one of those periods in the history of our own nation; and as we look back upon it, one of the first thoughts which must suggest itself to a believer in Providence is admiration at the wonderful manner in which that fierce contest, which brought into view and enlisted all the best and all the worst feelings of men, finally operated so mysteriously for good, in spite of the misery of so many years of internal convulsion. The effect of the evil was temporary; the good was permanent.

It is fortunate for us that we are able to view that event across the chasm of two centuries. The interval has made so marked an alteration in our social state, that most of the questions which relate to it may be studied without the party feelings which they once called forth.

We are now able to look at that period as a whole, in its consequents as well as its antecedents, and so to apprehend its real nature. For, as in estimating the effect of a work of human architecture, we are compelled to retire to some distance from it, if we wish to understand the unity of the artist's conception; so, if we would view rightly the great deeds of God's natural providence, we must first reduce them to their true historical perspective; and hence it is that when we observe the religious service of this day we do so with a different feeling from that which marked its early institution. For its founders viewed the scene as actors or spectators; we know the events of it only by narration. Their execration was directed against the persons who were distinguished in that revolution, ours against their crimes. Their minds could see nothing in the whole period but the death of the martyr king; we, while we can lament that cruel act as much as they did, can view it as one event in a whole period; they could only see that it was wrong to put to death one of royal blood; we, it is to be hoped, have learned the broader lesson, that it is wrong to employ capital punishment for any merely *political* offence whatever. They instituted the service to deprecate a trespass on the divine right of monarchical institutions; we retain it to assert our conviction of the divine right of human government.

We shall, accordingly, employ our time more profitably on the present occasion if we rise from the consideration of this single event to the contemplation of a general fact of God's providential government; for thus, instead of reviving wornout controversies or exciting party-spirit, we shall bear away that deep and reverential feeling which arises whenever we are made to perceive, by the study of God's general laws, something of the majesty, and wisdom, and goodness of Him, who filleth all in all.

The great fact to which attention is now asked is this,— that God has been pleased so to order the structure and arrangements of society, that even the suffering of periods of internal national convulsion is compensated by consequent good, and that the selfishness of parties of which those revolutions are the effect and manifestation, is overruled for the advancement and general happiness of society. A law like this is no defence for revolution,—it is no apology for insurrection. It is merely an argument from Final Causes in behalf of the moral attributes of the Almighty. It is one example of the truth which the piety of Solomon expressed in the words, "The Lord hath made all things for himself; yea, even the wicked for the day of evil."

It is necessary, however, before we develop the proof of the assertion, to point out briefly its bearing upon the general argument.

No fact has been made more clear, from the discoveries of science, than the abundance of beneficent arrangements which exist in nature. Each science contributes examples to the collective argument. Yet, wonderful as are the proofs of Divine goodness drawn from obvious instances of beneficence, they are not so striking as those which arise from observing the system of compensations ordained in nature for the misery and evil which exist.

It gives me a noble idea of the Divine Being when, as I watch the stars which move in the evening sky, I conceive of them as sustained in their orderly course according to a few simple intelligible laws; but how vastly is my idea deepened, both concerning the majesty and beneficence of the Almighty, when, in reflecting on the disturbances which they are generating in one another's movements, and trembling to think of the universal catastrophe which in the depths of future times those disturbances may bring about, I see that the subtle results of calculation demonstrate that a system of compensations of amazing grandeur is at work, and that the laws which the Divine Being has impressed upon matter guarantee the stability of the systems which He has created!* Or when I restrict my view to phenomena of this earth, which seem not to speak of a God of love, what an idea do I obtain of the Divine goodness, when it is made apparent that even the volcano's fire, which seems only fraught with desolation, is made the instrument of replenishing the whole vegetable kingdom, and thus, indirectly, the family of man, with a supply of those material principles which are necessary for its continued support!† The great fact of the permission of physical evil is brought before the mind in a new aspect. That single apparent exception to the Divine goodness is seen to be marked by evidences of it; and the surprise which the discovery excites, renders the argument irresistible.

And though analogy would hardly warrant an expectation that the intractable phenomena of moral and social evil would yield to explanation as readily as those already cited of physical mischief, yet here also experience brings to light the existence of a system of laws which are as comprehensible and unchangeable as those which regulate the universe of brute matter. Now, of social phenomena there are two classes, which at first view seem fraught with woes without commensurate blessings, and which a sceptic might regard as an objection to the idea of the government of a merciful Creator,—viz., *external war* and *internal convulsion*. The argument might be applied to the former of these topics; and war could be shown, however immense an evil in itself, to have contributed to the progress of civiliza-

---

* Lagrange's "Problems," referred to before in the first three Sermons.

† See the concluding chapter on "The Final Causes of Volcanoes," in Dr. Daubeny's work on that subject.

tion, and the final result of most of the conquests which the world has witnessed, be proved to have been beneficial; but it is not so obvious that the same remark applies to the latter subject. It is, accordingly, my object to extend the argument on the Divine benevolence to embrace this class of phenomena. It is intended to assert that, irrespective of the particular deeds of the parties whose interests may operate in public disorders, the general effect of such convulsions in the order of Providence is not the misery which would be antecedently expected from them.

There are two principal respects in which Divine Providence overrules the misery of revolution, viz., in making it contribute to the *material* and the *moral* welfare of man; —the *material*, in the advancement of liberty and happiness; the *moral*, in the formation of national character and opinion.

The few moments of our present service will not admit of that induction of particular instances of revolutions, which would be the natural mode of establishing this assertion. It must suffice to indicate the mode in which these advantages are brought about, and to advert to some features in the history of the revolution to which our thoughts are this day turned, which will afford illustration and verification of them.

I. History seems to show two facts in relation to society as fundamental circumstances of its existence; first, that society is in a state of progress; and secondly, that it is a progress towards a condition of equality.*

Power and government are given to be used for the good

---

* This subject may be studied in Aristotle's "Politics," b. v; in Vico's "Scienza Nuova;" in Dr. Arnold's "Thucydides," vol. i, appendix 1; in Thirlwall's "Greece," vol. i, ch. 7; in Lucas's "English Prize Essay, at Oxford, 1845."

of others, and not for the benefit of the possessor; they are a trust, and not a right. If this thought could be felt and acted upon, revolution would be unknown; when it is forgotten and violated, a convulsion is the terrible means of reasserting it. Wherever we possess the history of a revolution we shall find that it has always arisen from an opposition to the unalterable fact, that society is in a state of growth to maturity, and that, accordingly, institutions which are a blessing to one age may, after a lapse of time, cease to work for good; that power which is wisely withheld from men unfit to use it, becomes the right of people who are, by education or circumstances, fitted for its possession. Nor is it an objection to this doctrine that the point at which power ought to be conceded cannot be precisely laid down; for this is only stating a difficulty which is common to all sciences, which, like that of government, consider only probable evidence, and which, after exhibiting a principle, commit the execution of its details to the agent as part of his moral trial.

If, then, there be a truth in this principle, it is clear that revolution is made in the hands of Providence a good, if it is the means of developing that progressive growth in society which is its ineffaceable property. The assertion does not involve the propriety of revolution. The end may be good, the means which men employ for its attainment may be indefensible; it only postulates that through the means, whether right or wrong, the great end of public liberty and consequent happiness is brought about, and the beneficence of the Almighty thus shown in constructing society in such a manner that even the severity of revolution ministers to the permanent good of man; that the Lord hath made all things for himself, even the wicked for the day of evil.

If we apply this principle to the case of our own revolution we shall remark an illustration of its truth.

In order that we may see how this particular case is an instance of the law of progress just stated, we should notice that it assumed the form not so much of a contest for new rights, as of a protest against the infringement of those which already existed. This is a feature common to all the struggles of that age in European states, and forms a sufficiently marked contrast with those which we see exemplified in the ancient world. The cause is that, when about the beginning of the sixteenth century the several European nations, for the first time, began to act in their international relations as united kingdoms, each of them separately possessed in itself the reality of present and the prospect of future liberty. It was the heritage handed down from the past, the mixed result of the free institutions which were introduced by the barbarians who overran the Roman empire, and the municipal and social institutions which Rome had communicated to the world, and which survived the nation which had established them.* But at the time of which we speak, this liberty was to be put to a severer trial than it had ever known. It had outlived the incursions of barbarism; it was now to suffer danger from the very advance of civilization. For the circumstance which then enabled the different nations of Europe to take their position in the European system was, that they had each lately, for the first time, become consolidated into united kingdoms. The fragments of civilization had existed in the middle ages; but the attempts of great minds, such as Charlemagne and Hildebrand, to reunite those frag-

---

* See Guizot's "Hist. of Civilization," vol. i, ch. 6, 10; and Sir J. Stephen's "Lect. on Hist. of France," vol. i, ch. 3, 4.

ments, had been baffled.* Nature, however, in due time brought about that which they were unable to effect, first by joining the scattered elements of each state into one nation, and then by cementing the various nations of Europe into one confederation by the common principles of a states-system and law of nations.

It was in this centralization of power that the peril to public liberty in that age arose. The legislative and judicial functions were so arranged as to be a sufficient guarantee of liberty; but they were now, for the first time, in danger of succumbing to the executive. Hence the form which the struggles for public liberty took was that of the defence of the ancient constitutions. That this danger extended to England also at that period none will be prepared to deny. There may be differences of opinion as to the amount of opposition which the measures of the administration justly provoked. Some, for example,† may think that liberty was sufficiently secured by the first session of the Long Parliament in the spring of 1641, without the institution of an ex post-facto law to punish the supposed public offenders; others‡ may judge that further guarantees were required in dealing with a government which had flagrantly violated its own promises given to the Petition of Right twelve years previously; but whichever view be taken, there is not an Englishman living who will deny the peril in which public liberty was placed in the early part of the reign of Charles I by the plans of his counsellors. If proof were asked, it may be found in the

---

* Guizot's "Hist. of Civilization," vol. i, ch. 3, 6, 10.

† As Hallam, in his "Constitutional History."

‡ As Macaulay, in his review of Hallam's "Constit. Hist.," Essays, vol. i.

private letters of that acute statesman* who directed for many years the king's counsels. We there find that, enamoured with the beauty of the principle of centralization as carried out in the preceding century in Spain, and in his own day by the cardinal statesman who directed the counsels of France, he had planned schemes for increasing the power of the executive until the government of England should be assimilated to those of absolute kingdoms. Whether the success of such a scheme would have been for the good of this country will be seen if you further take into account the probability which existed at that time (a probability which subsequent events have made a fact) that England would rise in the scale of European kingdoms, and would consequently be compelled to establish a standing army.† For this circumstance was now about to give, for the first time, a power to the Prince to carry out his will in spite of the will of the nation. And the history of France may show us what would have been the effect in this country, if the providence of God had not by the great revolution prevented it. In that kingdom the ablest, not merely of her many able kings, but one of the most talented princes who ever adorned a throne, succeeded in crushing the liberties of his country, not from the low and selfish motive of his own gratification, but under the guidance of the enlightened Colbert,‡ from a mistaken view of providing for the general happiness of his people; and the consequence was seen soon after his death. He had caused all power to depend upon the character of the king. His

---

* The references to "Wentworth's Correspondence" are given in Macaulay's "Hist. of England," vol. i, ch. 12.

† See Macaulay's "Hist. of England," vol. i, ch. 1.

‡ For Colbert's administration, see Sir J. Stephen's "Lect. on Hist. of France," vol. ii, ch. 22.

successor, a weak and effeminate prince, was unable to carry out his great ideas. The unreality of the system was at once revealed to view, and the dreadful acts of the first French revolution were the protest of the people against the abuses of that system of government. If we could guarantee the perpetuation of a race of kings possessed of perfect intellectual and moral qualities, we could intrust to them absolute power: but such a property is confined to God's government; it exists not in man's. And, therefore, as we cannot insure the perfect wisdom, or the entire goodness of a race of governors, we establish public liberty on the basis of a Constitution, *i. e.*, we surround all those who are invested with power with such artificial regulations and conditions as may allow them full scope for acting for the good of the nation, but may bring them to a stand at once if either their judgment or their heart betray them.

We believe then that at the beginning of the reign of Charles I, England was in the great peril of passing from a constitutional to an absolute form of government, and we assert that the effect of the revolution which followed, was to guarantee and to assure the continuance of the Constitutional power. We are fully alive to the faults and sins of those who conducted the revolution; we do not even attempt to defend that revolution itself; it is unnecessary to the present argument to do so. We speak not of man but of God; not of human acts but of the Divine general plan; and we assert with confidence that any one who looks with pride on the British Constitution, and believes that as a whole it contains a surer guarantee for public liberty than any form of government which the world has ever known, must feel that whatever may have been the temporary evil of the revolution, the establishment of that Constitution on a sure basis was a permanent unspeakable good; we affirm that the providence of God, in permitting great periods of public misery, is seen nevertheless to compensate for them by lasting good. The evil falls on a single generation; the blessings become the everlasting inheritance of posterity. We assert that we have justified the utterance of Solomon, "The Lord hath made all things for himself; yea, even the wicked for the day of evil."

Yet it may be thought that our attention is called to-day, not so much to the revolution as to its crimes, and more especially to that act of malice, the spiteful vengeance taken on a helpless prince by men too mean to be generous, too selfish to be just. This aspect of the revolution contributes a most important point in the argument to the Divine beneficence. For any one who studies the history of that time must marvel how it was, that the country ever could free itself from the men into whose hands the government had fallen. And the same remark is true of all such periods in other nations, as well as in that of our own country. Suppose an individual in the ancient time attempting in the midst of some period of anarchy, such as the civil wars of the Roman Triumvirate, to forecast the future of Roman history; or during the great French revolution, wondering, as the faubourgs of the capital poured out band after band of revolutionists, each more fierce than its predecessors, how society was ever to deliver itself from the tyrants which itself had raised, and you will realize the beneficence of that law of Providence which enables society to rectify itself. At such times, the first leaders of the revolution, men of patriotic temper and high principle, succumb to impostors who come with unreal schemes of political change, and with dark and selfish plans for their own aggrandizement. Society is redissolved into its primitive elements. Law and moral power give place to physical. And yet the day of deliverance comes. If it were not so, a convulsion would be an unmixed woe, a vial of wrath to the age which suffers it, and a curse to posterity. But there is in man a tendency to preserve as well as to destroy, an instinct against anarchy as well as against tyranny. And by this principle, a beneficent Providence enables society to free itself from its own excesses, and to secure the real blessings which it originally coveted, without permanent submission to the evils which attended their attainment.

The civil war in our own history supplies abundant evidence of the truth of these remarks. Omitting the many wild schemes of political change which were then fermenting in society, we may allude to the danger which menaced the established Church of this country. The Constitution of England did not undergo a greater peril from the absolutism of Wentworth, than the faith, and art, and institutions of its Church underwent from the fanaticism of the Puritans. While we must always admire the learning and personal excellence of many of the Puritan clergy, and their steady and praiseworthy (though often excessive) attachment to the Protestant faith, yet we cannot but feel that the success of their principles would have involved the destruction of some articles of faith, and some Apostolical institutions which our Church holds most dear; and would especially, by the addition of harsh tests and dogmatic creeds, have extinguished that breadth and comprehensiveness of character which has always formed the glory of the English Church, and would have suppressed the freedom of thought which must ever be the only permanent safeguard of theological truth. And so the character of Charles I must always carry an interest to the hearts of churchmen, for one of the chief features of his mind was his marked attachment to the English Church. There is even something chivalrous in the manner in which he and the royal party defended the Church, as dearly as the throne. Those heroic men died, and their last moments were saddened by the thought that their cause was lost; but they died not in vain. They fell defending the breach through which the enemy was entering the fortress; but men, as they saw them fall, themselves regained their courage, and rallied round the standard which they had died defending. And so the cause of order again revived, and the moral again triumphed over the physical. Or if you prefer to explain in some other manner the play of human passion through which society was enabled to reap the solid benefits of the revolution without becoming the permanent victim of its excesses, you must at least confess that it is the beneficence of the Almighty which has given society the power to preserve itself. Human forces seem to hush the chaos of anarchy, but it is the Lord who says, Peace; be still. The storm passes away, the calm again returns. It is the Lord who "hath made all things for himself; yea, even the wicked for the day of evil."

II. We stated that it was not merely in the *material*, but in the *moral* benefit of national convulsions that the beneficence of Providence can be shown.

The moral benefits are to be found either in the discipline which is administered by them, or in the lessons taught. As we distinguish between the training and the instruction of an individual mind, so also the same difference exists between the gradual and almost unobserved formation of national habits, and the opinions which constitute the popular and conscious states of belief.

The discipline resulting from times of convulsion is necessarily that of suffering. The tendency of suffering to produce good, as applied either to an individual or a nation, is an allowed fact. Yet it may occur to the mind that the suffering of revolution, being connected with party strife, cannot administer the same blessings as in instances where it arises from external sources. It is obvious how an event, such as a public war, may unite a people in a common purpose and a common sorrow, and tie them by a principle of patriotism in the bond of one brotherhood. But it is not so clear how good results can follow when the suffering arises from strife and division. There is indeed truth in this remark. Yet it is the marvellous property of pain that it possesses so comprehensive a power, and so appeals to every deep susceptibility of our nature, that it lays bare some feelings which operate for moral discipline, even in a case like this, which at first sight is so unpromising. For suffering brings home to man a deep practical sense of his own insufficiency, and his dependence on a superior power. There are two principal feelings in the mind, the consciousness of dependence and the consciousness of power; the latter prompts to action, the former to piety. Suffering acts upon the latter; it awakens in us the consciousness of needing help which we cannot supply. Like Manasseh in his dungeon, we are then ready to call on the Lord God of our Fathers. And suffering also draws out some special virtues. Though in mean characters it increases only selfishness and cruelty, yet in others it calls out courage and generous devotion. It singularly happens, too, that a selfish principle is added in cases of revolution to assist it; for the spirit of party excites a feeling of sympathy with those who are ranged on the same side.

And, in fact, history might be brought in to confirm the benefits which theory establishes.

But suffering serves not only to discipline, but to instruct; for linked as it is with the notion of sin as its cause, it excites a deep practical hatred of the crimes which have produced the suffering. And, connected with this, it has a tendency to awaken attention and induce inquiry concerning both the cause and cure of national convulsions. How many lessons might thus be gathered from our own revolution; how many precious maxims of political wisdom or moral guidance might be its warning to posterity! Nay, even if it should be found that men through carelessness have neglected to gather those lessons, yet if it be shown that instruction is the natural though not the necessary result of suffering, our argument equally proves that even in the pain attending on public convulsions, the beneficence of the Almighty is manifested.

We have now completed our notice of some points in relation to public convulsions, in which the goodness of God is seen in bringing permanent good out of temporary evil. And you must be again reminded that the argument is irrespective of the question of the propriety of any particular revolution, or even of revolution in the abstract. We assert that the Almighty has been pleased so to govern society that he rescues it from the effect (it may be) of its own follies; that he evolves good where we have only a right to expect evil; that "He hath made all things for himself; yea, even the wicked for the day of evil."

In conclusion, we may bear away two valuable lessons:

1. The consideration of the benevolence of the Almighty ought to give us confidence in the contemplation of the future prospects of the nation or the world.

At all times, earnest men feel so bitterly the disappointment of their hopes that they are ready to despair of the prospects of mankind. Their experience teaches them that the strongholds of evil bid defiance to their attacks,—that their efforts for social or political amelioration meet only with defeat.*

And if we look at the material and moral aspect of the world, there are many circumstances to suggest thoughts of deep sadness. Eighteen hundred years have passed since the glad tidings were proclaimed that a Saviour had come; and still three-fourths of the population of the world have never yet heard of His name. Or if we restrict our view to our own land, which seems in a special manner to be the home of civilization and piety, what a terrible state of society is revealed to us by that single statistic,† that one-half of our people never enter a place of worship! What prospect is there of reclaiming, even for civilization itself, the masses crowded by thousands in the hearts of our large towns, who have thrown away virtue, humanity, and religion? It is not surprising that earnest men should look with gloom on the future.

But the subject which we have been considering may afford some ground of consolation. Though the prospect be really dark,—though the shadows of night seem closing in upon us, yet at evening time it shall be light. There is One above, whose eye is not unmindful of this world's history, whose blessed Son "tasted death for every man," —who "willeth that all men should be saved." And He so manages this world's course that He will evolve, by His general laws, good out of evil. He rules not in the world

---

* A paragraph of the original Sermon is here omitted which referred to circumstances of war and recent disease, now happily past.

† A fact of the census of 1851.

merely of blind unconscious matter, but in the actions of thinking responsible man. The past of the world declares that "He hath made all things for Himself; yea, even the wicked for the day of evil." And if we could take our stand on some eminence, and trace forward, in the depths of the future, the issue of this world's stream of time, we should see it swallowed up in the ocean of the Divine attributes. We should take up the song of Seraphim, and proclaim, that not the heaven merely, but the earth also, is full of God's glory.

2. There is a second lesson, which is a very practical one, taught us by this subject, viz., to ask ourselves whether we are severally, by our conduct, co-operating in carrying out the plan of Providence, or helping to thwart it. The end of the Divine government is goodness. If we are aiming at the same result, we are filling our proper sphere in the world; if our hearts are full of selfishness toward man and disobedience toward God, we are doing our part to thwart the great end of the Divine administration. How blessed a privilege it is to think that we have the opportunity, each one of us, of doing our share—aye, though it be the very smallest—to co-operate with God! How much might we do for the world by our efforts if we were to use our opportunities of battling with evil! Let us understand that the way to regenerate others is first to renovate ourselves. If we would be the means of doing something for God and for goodness, we must fix well in our minds that we are not to reflect the religious tone of the world, but to introduce into the world the elements of goodness which it does not possess. It is God the Holy Ghost who alone can give them to us. It is He alone who can impart to us that religious earnestness which shall make us active for His honor and for man's welfare. The chief mode of

obtaining this Divine help is by cultivating a habit of communion with God by private prayer. This is the secret of any good man's life, whether it be one of action or of suffering. It was so with the royal prisoner. As his troubles thickened round him, and he could find no help from friends and no mercy from foes, he was wont to betake himself to God for consolation; and hence arose that placid, confessor-like spirit with which he bore the sorrows of his closing days, and that calm and heroic fortitude with which he met the terrors of the hour of death. And we may be certain that, whatever be our earthly lot, however humble or however great, whether spent in happiness or marked with sorrow, if we would secure our own usefulness in life, and our own safety in death and judgment, we must follow such an example of piety, and learn to find a friend in God. For then we shall be as polished shafts in the Almighty's quiver in the great battle of good against evil; we shall indeed find that evil will give way before us; and, in the blessed consciousness of a life not spent altogether in vain, we shall realize in their fulness the words of Solomon, "The Lord hath made all things for Himself; yea, even the wicked for the day of evil."

NOTE,

On the Scene of the Execution of Charles I.

This seems not to be an unfit place to notice some facts in reference to the scene of the king's execution, which have been drawn from old engravings and maps, still preserved in the Chapel Royal at Whitehall. The king was executed in front of the middle window of the present Chapel Royal, on the side facing the present street, and not, as is often supposed, on the other side. At that time, instead of the streets and gardens which now lie around, an old brick palace existed, not unlike parts of the present one of St. James's. Its outlying quadrangles and buildings stretched as far north as the present Scotland Yard, while one large quadrangle, containing the royal garden, lay immediately to the back of the Chapel Royal; on the side of which quadrangle, next the river, stood the royal apartments. The street which now runs in front of the chapel was about half its present width; a guard-house stood in front of the present Horse Guards, while immediately in front of the chapel was a tilting-ground; and a few yards to the south of it, *i. e.*, in the direction of Westminster Abbey, a brick archway spanned the street, similar to that which now forms the principal entrance to the Palace of St. James. The banqueting-hall which forms the present Chapel Royal is the only portion ever completed of a grand design of James I for rebuilding the palace. The older portion of the palace was destroyed by fire in the time of William III, and the banqueting-hall was converted into a chapel by George I. On the day of the execution, Charles I was brought (about ten o'clock in the morning) from the Palace of St. James across St. James's Park, and was conducted over the archway, which has been above described. He then spent nearly three hours in worship, probably in a small chapel which then lay adjacent to the archway to the southeast of the present Chapel Royal; and, after his devotions, was conducted through the interior of the present chapel to the scaffold. It is doubtful whether he passed through one of its windows on to the scaffold, or was led completely through it to a portion (now destroyed) of the palace which then stood a little to the north of the present chapel, and thence led to the scaffold; but that the position of the scaffold was in front of the present building there can be no doubt.

www.ingramcontent.com/pod-product-compliance
Lightning Source LLC
Chambersburg PA
CBHW020804230426
43666CB00007B/839